INDUSTRIAL PROJECT MANAGEMENT

Concepts, Tools, and Techniques

ADEDEJI BADIRU
ABIDEMI BADIRU
ADETOKUNBOH BADIRU

CRC Press
Taylor & Francis Group
Boca Raton London New York

CRC Press is an imprint of the
Taylor & Francis Group, an **informa** business

CRC Press
Taylor & Francis Group
6000 Broken Sound Parkway NW, Suite 300
Boca Raton, FL 33487-2742

© 2008 by Taylor & Francis Group, LLC
CRC Press is an imprint of Taylor & Francis Group, an Informa business

First issued in paperback 2019

No claim to original U.S. Government works

ISBN-13: 978-0-367-45297-1 (pbk)
ISBN-13: 978-0-8493-8773-9 (hbk)

Visit the Taylor & Francis Web site at
http://www.taylorandfrancis.com

and the CRC Press Web site at
http://www.crcpress.com

Library of Congress Cataloging-in-Publication Data

Badiru, Adedeji Bodunde, 1952-
Industrial project management : concepts, tools, and techniques / Adedeji Badiru, Abidemi Badiru, and Adetokunboh Badiru.
p. cm.
Includes bibliographical references and index.
ISBN-13: 978-0-8493-8773-9 (alk. paper)
ISBN-10: 0-8493-8773-6 (alk. paper)
1. Production management. 2. Project management. I. Badiru, Abidemi. II. Badiru, Adetokunboh. III. Title.

TS155.B2175 2007
658.4'04--dc22   2007006778

INDUSTRIAL PROJECT MANAGEMENT

Concepts, Tools, and Techniques

Industrial Innovation Series
Adedeji B. Badiru
The University of Tennessee, Knoxville, Tennessee

Published Titles

Computational Economic Analysis for Engineering and Industry
Adedeji B. Badiru & Olufemi A. Omitaomu

Handbook of Industrial and Systems Engineering
Adedeji B. Badiru

Industrial Project Management: Concepts, Tools, and Techniques
Adedeji B. Badiru, Abidemi Badiru, and Adetokunboh Badiru

Techonomics: The Theory of Industrial Evolution
H. Lee Martin

Forthcoming Titles

Process Optimization for Industrial Quality Improvement
Ekepre Charles-Owaba and Adedeji B. Badiru

Dedication

Dedicated to the memory of Omolade Badiru
The bud that never got to bloom, but whose spirit lives on

Contents

Preface

Industry represents the pulse of economic development of any nation. Successful industrial project management thus holds a key position in advancing regional and national development. Project management is the process of managing, allocating, and timing resources to achieve a given goal in an efficient and expeditious manner. The objectives that constitute the specified goal may be in terms of time, costs, or technical results. Projects can range from the very simple to the very complex. Owing to its expanding utility and relevance, project management has emerged as a separate body of knowledge that is embraced by various disciplines ranging from engineering and business to social services. Project management techniques are widely used in many endeavors, including construction management, banking, manufacturing, engineering management, marketing, health care delivery systems, transportation, research and development, defense, and public services. The application of project management is particularly of high value in industrial enterprises. In today's fast-changing and highly competitive global market, every industrial enterprise is constantly striving to get ahead. Integrative project management offers one avenue to achieve that goal.

Project management represents an excellent basis for integrating various management techniques such as statistics, operations research, Six Sigma, computer simulation, and so on. The purpose of this book is to present an integrated approach to project management for industrial applications. The integrated approach covers the concepts, tools, and techniques (both new and tested) of project management. The elements of the project management body of knowledge provide a unifying platform for the topics covered in the book. The book also contains a project-oriented chapter on Lean Six Sigma. The chapters of the book are

Chapter 1: Characteristics of Industrial Projects
Chapter 2: Principles of Project Management
Chapter 3: Time and Schedule Management
Chapter 4: Project Duration Diagnostics
Chapter 5: Schedule Compression Techniques
Chapter 6: Resource Analysis and Management
Chapter 7: Techniques for Project Forecasting
Chapter 8: Six Sigma and Lean Project Management
Chapter 9: Project Risk Analysis
Chapter 10: Project Economic Analysis
Chapter 11: Industrial Project Management Case Studies

Appendix A of the book presents project terms and definitions, and Appendix B presents project acronyms. The premise of the book is that both simple and complex industrial projects can be managed better if an integrated approach is utilized. The integrated approach in the book covers managerial principles and analytical techniques. The book presents tools and techniques for mitigating the adverse effects of the typical constraints of time, cost, and performance in any project.

This book is intended to serve as a reference for project planners, designers, and managers; as a guidebook for industrial consultants; as a textbook resource for students and teachers; as a supplementary reading for practicing engineers; and as a handbook for project operators. It will appeal a great deal to practitioners and consultants because of its practical orientation.

Adedeji Badiru
Abi Badiru
Ade Badiru

Acknowledgments

Adedeji Badiru: I thank the many friends and colleagues who continue to support and encourage me in my cycles of writing projects. I thank my wife, Iswat, for continuing to tolerate my love affairs with new writing projects. I thank Christine Tidwell, Jeanette Myers, and Louise Sexton for their administrative and technical word processing support for the manuscript preparation. Christine diligently prepared the several figures and charts contained in the book. I thank Em Chitty Turner for bringing her usual technical editing skills to bear on this project. I also thank the production staff of CRC Press of Taylor and Francis for the excellent job of sorting through the editorial maze of the raw manuscript. We thank F. B. Odunayo, Managing Director of Honeywell Flour Mills, Lagos, Nigeria, a foremost international industrialist, whose initial insightful comments and inquiry mind led to the idea of writing this book.

Abi Badiru: I extend my thanks to my extended circle of friends from different parts of the world. I extend appreciation to my present and former colleagues from Kraft Foods and ConAgra Foods.

Ade Badiru: Many thanks and love to my wife, Deanna Badiru, whose continual love and support inspire me in all my professional endeavors.

Authors

Professor Adedeji Badiru is the head of Systems & Engineering Management at the Air Force Institute of Technology. He was previously the department head of Industrial & Information Engineering at the University of Tennessee in Knoxville, and professor of industrial engineering and dean of University College at the University of Oklahoma. He is a registered professional engineer. He is a fellow of the Institute of Industrial Engineers and a Fellow of the Nigerian Academy of Engineering. He holds a BS in industrial engineering, an MS in mathematics, and an MS in industrial engineering from Tennessee Technological University, and PhD in industrial engineering from the University of Central Florida. He is the author of several books and technical journal articles.

Sikiratu (Abi) Badiru was born in Lagos, Nigeria, and moved to the United States at the age of six. She grew up in Norman, Oklahoma, and graduated from the University of Oklahoma with a BS in chemical engineering in 1999. She also received an MBA from the University of Texas, Dallas, in 2003 with a specialization in organizational behavior. After completing her BS she went to work for Kraft Foods as a project engineer for 5 years. She currently works as a project engineer/manager for ConAgra Foods in Omaha, Nebraska. Abi resides in Bellevue, Nebraska.

Ibrahim (Ade) Badiru grew up in Norman, Oklahoma, and graduated Summa Cum Laude from the University of Oklahoma with a BS in mechanical engineering in 2000. He also received an MS in mechanical engineering from the University of Michigan in 2002 with a research focus in the area of design optimization. He currently works as a Vehicle Dynamics Engineering Specialist for General Motors in Milford, Michigan. During his time with General Motors he has held a variety of project engineering positions within the Product Development Group and has served as a consultant to the General Motors automotive racing organization. He also serves as a board member to the University of Oklahoma College of Engineering. Ade lives with his wife Deanna in Northville, Michigan.

1 Characteristics of Industrial Projects

IMPORTANCE OF INDUSTRIAL PROJECTS

Industry represents the pulse of economic development of any nation. The goods and services provided by industry directly influence the social, political, economic, and cultural structures of any population. Thus, successful industrial project management holds a key position in advancing local, regional, and national development. A community that cannot institute and sustain industrial vitality will eventually become politically delinquent and economically retarded. Project management is the process of managing, allocating, and timing resources to achieve a given goal in an efficient and expeditious manner. The intrinsic benefits of this definition are even more pronounced in fast-paced and globally influenced industrial projects.

TIME–COST–RESULT GOALS OF INDUSTRY

The objectives that constitute industrial project goals may be in terms of time, costs, or technical results. Projects can range from the very simple to the very complex. Owing to its expanding utility and relevance, project management has emerged as a separate body of knowledge that is embraced by various disciplines ranging from engineering and business to social services. Project management techniques are widely used in many endeavors, including construction management, banking, manufacturing, engineering management, marketing, health care delivery systems, transportation, research and development, defense, and public services. The application of project management is particularly of high value in industrial enterprises. In today's fast-changing and highly competitive global market, every industrial enterprise is constantly striving to get ahead. Integrative project management offers one avenue to achieve that goal.

LASTING LEGACY OF PROJECT MANAGEMENT

Project management has had more direct impacts on human development than any other single discipline of study in the history of the world. From the time of ancient history and Mesopotamia's early development to the modern times, acts of project management have brought to bear on human accomplishments. Early examples include the construction of Stonehenge in England, the erection of the Pyramids, and the development of the notable Wonders of the World. The ancient projects using gears and pulleys required extreme preparation, labor coordination, and cooperation. Although there was no formal discipline of project management in those ancient times, the processes of planning, organizing, scheduling, and control, no doubt, were used in accomplishing those feats. In spite of its long-standing benefits, it was only in the past few years that project management has emerged as a formal discipline; and it is now being globally recognized. The Project Management Institute has an envisioned goal that states, "Worldwide, organizations will embrace, value, and utilize project management and attribute their success to it." This vision is already being broadly realized. This is evidenced by the rapid growth in project management professional memberships around the world. Interest in the discipline is growing rapidly around the world—in Europe, Asia, North America, South America, the Far East, the Caribbean, Africa, and so on. There is no single country that can claim not to be touched daily by the impact of project management processes.

ELEMENTS OF INDUSTRIAL OPERATIONS

Industrial development is one primary path to achieving national economic development. So, industry is very vital to the development of any nation. Historical accounts abound on how the industrial revolution had a profound effect on world development. A sustainable industrial development can positively impact the political, economic, cultural, and social balance in a community. In order to achieve and sustain industrial development, both the technical and managerial aspects of industrial projects must come into play. This book focuses on the integration of managerial approaches and analytical techniques to improve the planning, scheduling, and execution of industrial projects.

The primary goal of any industry is to plan operations ahead and allocate resources appropriately to improve industrial project efficiency, effectiveness, and productivity while reducing production waste (Lean) and improving product quality (Six Sigma). Using a formal project management approach makes it possible to achieve this goal. For projects to be effectively managed in an industrial system, managers and analysts must understand the industrial operating environment. Any high-tech industrial project is a complex undertaking that crosses diverse areas of endeavors. Both technical and organizational issues must be addressed in order to avoid system-wide project failures. This chapter covers the building blocks essential for the application of project management to industrial operations. The contents of this and the subsequent chapters will enable the project analyst to accomplish the following learning objectives:

- Understand the basic steps and components of project management.
- Learn best-practices approach to project planning, organizing, scheduling, and control.
- Use case examples as the basis for understanding "what went wrong" and how develop sustainable project solutions.
- Learn how to develop project scope and develop a project charter.
- Using planning as the roadmap toward project success.
- Create cohesive project teams using the Triple C model of communication, cooperation, and coordination.
- Develop project work breakdown structure.
- Use a mix of qualitative and quantitative techniques to enhance project management.

- Develop compromise or tradeoff strategies for cost, schedule, and performance constraints.

Manufacturing is the process of creating a product by processing raw materials from an initial point through to the end product. It encompasses several functions that must be strategically planned, organized, scheduled, controlled, and terminated. A manufacturing cycle includes such functions as forecasting, inventory control, process planning, machine sequencing, quality control, decision analysis, production planning, cost analysis, process control, facility layout, work analysis, and a host of others. All of these are functions that fall within the process of planning, organizing, scheduling, and control cycles of project management. Industrial projects can be characterized by a combination of the following attributes:

- Large external stakeholders, customers, owners
- Internal stakeholders
- Short product life cycle (in high-tech industries)
- Variable investment sources
- Narrow margins for success.

As with all projects, industrial projects are subject to three basic constraints of time, cost, and performance as illustrated in Figure 1.1. Any other constraint in the project environment will somehow fall under one of these three constraints. Several factors lurk behind the screen of the triple constraints. Issues such as workforce capability, operating tools, and process structure impinge on the project's ability to be delivered on time, within budget, and in line with performance expectations.

Industrial operations are predicated on strategic operations, which utilize high-tech tools, knowledge workers, and complex processes. Consequently, project management in an industrial operation implies the management of people, process, and technology, as shown in Figure 1.2, to satisfy the triple constraints.

While the proliferation of technology in industry has led to a loss of jobs, it has also led to the creation of new types of jobs, and so the coupling of technology and manufacturing has spawned a need for retraining of workers and realignment of functions. Even though high technology is sometimes blamed for stifling creativity and restricting traditional personal workmanship, it has also been credited with fostering *industrial innovation*.

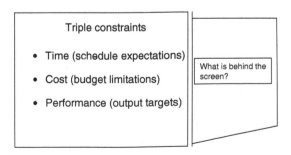

FIGURE 1.1 Triple constraints on projects.

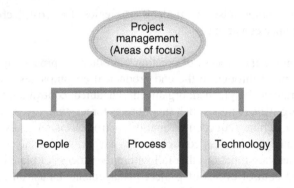

FIGURE 1.2   Focus on people, process, and technology.

This requires new management approaches. Effectively managing industrial technology requires project management skills on the part of management, employees, and clients in order to ensure the successful design, development, production, transfer, introduction, and implementation of various forms of technology to generate products and or services. Innovative applications of new and existing management techniques are needed to address the rapidly changing nature of industrial operations. Project management approaches are at the forefront of such applications.

DEPENDENCY ON HUMAN CAPITAL

In spite of the increasing proliferation of automation in industry, human capital still holds a major role in accomplishing industrial output. Investment in human resource assets should be a primary focus of any organization's project efforts. The success of the Toyota production system is not due to any magical properties of the approach, but rather due to the consistency, persistence, and dedication of the humans who apply the Toyota approach to all their industrial projects. This cannot be achieved without giving something (e.g., operator training, technology tools, and doable process) to obtain desired outputs. Recalling the cliché of "nothing from nothing is nothing," as illustrated graphically in the following figure, industrial organizations should invest in their human capital in order to maximize project output. Figure 1.3 shows the central role of people in the various aspects of an industrial system.

Nothing – Nothing = Nothing

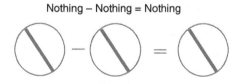

GLOBAL INDUSTRIAL COMPETITION

Many North American manufacturers cannot compete globally on the basis of labor cost, where improvement efforts are often directed. The competitive advantage for

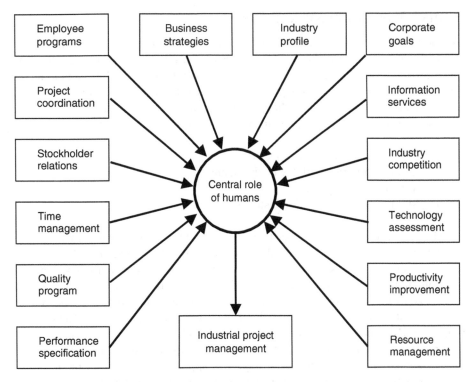

FIGURE 1.3   Role of human decisions in industrial projects.

many manufacturers will come from appropriate infusion of technology into the manufacturing enterprise. Strategic research, development, and implementation of technological innovations will give manufacturers the edge needed to successfully compete globally. In spite of the many decades of lamenting about the future of manufacturing, very little has been accomplished in terms of global competitiveness. Part of the problem is the absence of a unified project management approach. Managing global and distributed production teams requires a fundamental project approach.

One valid industrial proposition is the need to pursue more integrative linkages of technical issues of production and the operational platforms available in industry. Many concepts have been advanced on how to bridge the existing gaps. But what is missing appears to be a pragmatic project-oriented road map that will create a unified goal that adequately, mutually, and concurrently addresses the profit-oriented focus of practitioners in industry and the knowledge-oriented pursuits of researchers in academia. The problems embody both scientific and management issues. Many researchers have not spent adequate time in industry to fully appreciate the operational constraints of industry. Hence, there is often a disconnection between what research dictates and what industry practice requires. An essential need is the development of an industrial project road map. Two aspects that are frequently ignored in industrial project implementations involve human factors and ergonomics parameters of the work environment. A project management approach facilitates an appreciation of this crucial component of industrial projects.

System's View of Industrial Projects

An industrial system is a collection of interrelated elements brought together to achieve a specific objective of meeting product or service goals. In a management context, the purposes of a system are to develop and manage operational procedures and to facilitate an effective decision-making process. A systems approach is particularly essential for contemporary manufacturing because of the various factors that interact. Four of the major desired characteristics of an industrial project system include:

1. Possession of a definite objective
2. Ability to interact with the environment
3. Ability to self-regulate
4. Ability to carry out self-adjustment.

The various elements (or subsystems) of a system act concurrently, in a separate but interrelated fashion, to achieve the common goal. This synergism helps to expedite the decision process and to enhance the effectiveness of decisions. The supporting commitments from other subsystems of the organization serve to counterbalance the weaknesses of a given subsystem. Thus, the overall effectiveness of the system will be greater than the sum of the individual efforts of the subsystems. The increasing complexity and globalization of industrial operations make the systems approach essential. The classic approach to the decision-making process follows rigid lines of organizational charts. By contrast, the systems approach considers all the information interactions necessary between the various elements of an organization. The industrial system has shifted considerably over the past decades as illustrated in Figure 1.4. The primary focus in the 1960s was on industrial efficiency. Today, we are concerned not only with globality, but also with nanoscale industrial production; and cyber-space consciousness is already making dominant inroads into every level of project operations.

Industrial Project System Integration

Any project can be viewed as a system of operations and activities. There are several major steps for successfully initiating, implementing, and managing a project system. Some of the steps are summarized as follows:

1. *Definition of Problem*: Define the problem using keywords that signify the importance of the problem to the overall organization. Prepare and announce the project scope and plan.
2. *Assignment of Personnel*: The project group and the respective tasks and responsibilities should be explicitly established.
3. *Initiation of the Project*: Arrange organizational meetings and project kickoff, during which a general approach to the project is announced.
4. *Development of System Prototype*: If applicable, develop a prototype system, test it, and learn more about the problem from the test results.

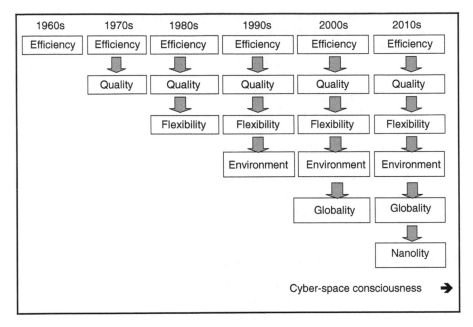

FIGURE 1.4 Evolution of industrial systems performance.

5. *Full-System Development*: Expand the prototype to a full system, evaluate the user interface, and incorporate user training facilities and documentation.
6. *System Verification*: Get experts and potential users involved, ensure that the system performs as designed, and debug the system as needed.
7. *System Validation*: Ensure that the system yields expected outputs. Validate the system by evaluating performance level.
8. *System Integration*: Implement the full system as planned, ensure the system can coexist with systems already in operation, and arrange for technology transfer to other projects.
9. *System Maintenance*: Arrange for continuing maintenance of the system. Update project system procedures as new information becomes available.
10. *Documentation*: Prepare disseminate documentation of system.

With increasing shortages of resources, more emphasis is placed on the sharing of resources, both physical and intellectual. It is through the integration of industrial systems that resource sharing may be most efficiently achieved. Systems integration may involve the physical integration of technical components, the objective integration of operations, the conceptual integration of management processes, or a combination of these. Systems integration involves the linking of components to form subsystems and the linking of subsystems to form composite systems within a single organization or across several organizations. Such integration facilitates the coordination of diverse technical and managerial efforts to enhance organizational functions, reduce cost, save energy, improve productivity, and maximize the utilization of resources. Because information and other resources are shared, it helps to ensure that components and subsystems operate synergistically to optimize the performance of the total system. Some important benefits of systems integration are as follows:

1. *Multiuser Integration*: This involves the use of a single component by separate subsystems to reduce both the initial cost and the operating cost during project life cycle.
2. *Resource Coordination*: This involves integrating the resource flows of two normally separate subsystems so that the flow of resources from one subsystem to the other minimizes the total resource requirements.
3. *Functional Integration*: This involves the restructuring of functions and the reintegration of subsystems to optimize costs when a new subsystem is introduced.

Resource Sharing for Systems Integration

Systems integration should cover both machines and people. Just as with physical systems, the supporting cooperative actions of personnel subsystems serve to counterbalance the weaknesses at certain points in the organization. The following is a representative list of possible subsystems. Figure 1.5 shows resource sharing plays a central role in project systems.

- Management
- Manufacturing
- Quality information
- Financial information
- Marketing information
- Inventory information
- Personnel information
- Production information
- Design and engineering
- Research and development
- Management information.

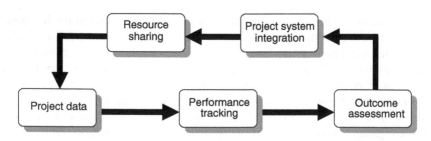

FIGURE 1.5   Resource-sharing linkages in project system.

Business Process Reengineering

Business process reengineering (BPR) is the redesign of business work processes and the implementation of the new design. This has emerged in recent years as a way to manage manufacturing functions. BPR calls for changes at three levels of the organization:

1. Enterprise-wide changes (driven by management initiatives)
2. Process-level improvement changes (driven by project teams)
3. Task-level changes (driven by individual workers).

Improvement in personnel skills and functions, technology, and in the process itself all contribute to the achievement of business process improvement. The factors that drive BPR are the needs for efficiency, quality, flexibility, and competitiveness. Traditional industrial processes operate in "blobs" (blurb) as depicted in Figure 1.6, whereby tasks are executed along fuzzy lines of responsibility. Although inputs and outputs may be lineated, within-the-box operations are often not easily tractable.

An alternative is to use a project system point-to-point lines of operations control as shown in Figure 1.7. Process inputs are clearly identifiable, the integrated outputs are observable, and the internal operations are clearly traceable. This has the advantage of the ability to trace points and sources of project problems.

CONTINUOUS PATHS OF IMPROVEMENT

If a traditional approach to industrial process improvement is followed, as shown in Figure 1.8, then alternate cycles of process improvement and degradation occur. This impedes overall potential to achieve the target. As an alternative, it is recommended

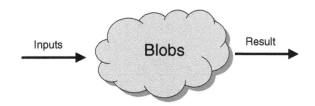

FIGURE 1.6 Traditional blobs of operation.

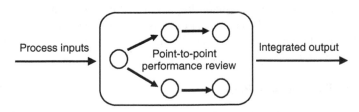

FIGURE 1.7 Point-to-point network of industrial tasks.

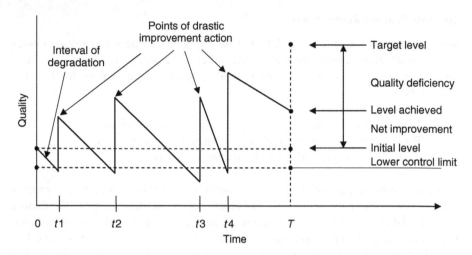

FIGURE 1.8   Undesirable traditional process improvement approach.

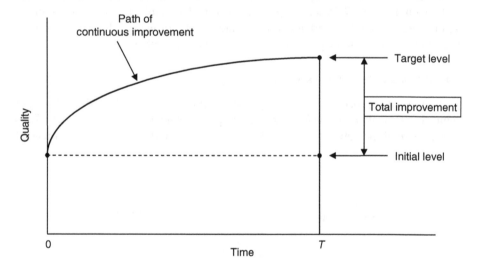

FIGURE 1.9   Desirable approach to industrial process improvement.

that industrial process improvement be pursued through incremental steps as depicted in Figure 1.9, whereby a continuous path is charted from the starting point to the target point. This has the advantage of lower cost, smoother operations, and greater potential to achieve production goals.

INDUSTRIAL PRODUCTION PLANNING

Production planning is the process of coordinating activities to get raw material stage to the finished-goods stage. It is the function of making sure that new designs are added as new products while old products are modified or discontinued. It involves setting up production objectives, allocating resources, and establishing standards and procedures to govern the production environment. The production-planning function is an iterative process that is reviewed and revised based on the state of the system. Some of the production tasks that may need to be coordinated as project activities include:

1. *Materials and Supplies*: Generation of reports showing materials and supplies assigned to and used by each production center.
2. *Labor Requirements*: Analysis of labor hours required for production operations.
3. *Overhead Allocation*: The distribution of overhead to production centers. An analysis is made of overhead allocated and overhead actually incurred for specific jobs.
4. *Job Control*: Tracking of job status and milestones.
5. *Job Transfer*: Routing of a job from one production center to another.
6. *Supply and Demand Trending*: Seasonality of certain products.
7. *Inventory Management*: Tracking of physical assets and resources of the organization.

PROJECT MODEL OF INDUSTRIAL PRODUCTION

The manufacturing enterprise is a project consisting of distinct production activities. In a large organization, the industrial system may be configured as a multiproject endeavor, and the components of the project may be managed as any conventional project. Figure 1.10 shows the typical components of an industrial enterprise organized as a project network of industrial.

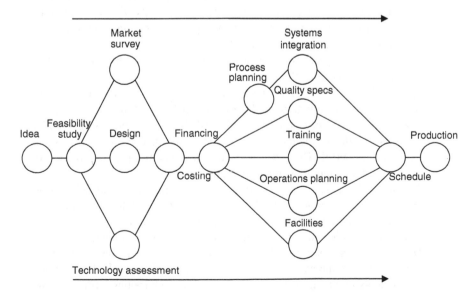

FIGURE 1.10 Manufacturing project network.

The industrial network starts with the conceptualization of a product. Some of the distinct tasks required for getting the product from the idea stage to the market include the following:

1. *Feasibility Study*: A study conducted to ascertain the practicality of the proposed product. The practicality is considered in terms of available technology, cost constraints, production process, labor skills availability, organizational goals, and market structure.
2. *Market Survey*: An analysis of what the market wants in terms of cost, product functionality, and comparative manufacturer reliability.
3. *Cost Estimation*: Development of cost figures for the various physical and qualitative components that go into obtaining the final product. These may include machines, inventory space, training, raw materials, transportation, advertising, design, customer service, labor wages, and so on.
4. *Technology Assessment*: An assessment of the current technological capabilities. This may involve questions such as: Is the technology proven and stable enough to sustain production operations? Is the technology affordable? What will be the impact of technology changes on production?
5. *Product Design and Development*: The development of a full-fledged design of the product based on the outcomes of the preceding tasks. Adaptability should be incorporated into the design so that future product changes can easily be accomplished as the technology changes.
6. *Financing*: The process of obtaining funds to complete the manufacturing project. Sources of funds may include top management (internal funding), external sponsors (e.g., government-backed projects), or contract awards (e.g., client-sponsored products). If the preceding tasks of feasibility study, market survey, cost estimation, and technology assessment are done with proper attention to details, the task of obtaining funds will be simplified.
7. *Process Planning*: The development of a plan for the manufacturing process, taking into consideration machines, tools, layouts, and raw materials. The level of process sophistication depends on the designed product configuration and availability of funds.
8. *Quality Specification*: The development of product quality specifications based on product functions and process capability. A process capability analysis may be very helpful in this task because a process capability analysis will determine the tolerance that a process can handle if it is statistically in control. The process, in this case, is a unique combination of machines, tools, methods, materials, and labor skills.
9. *Personnel Assignment and Training*: At this stage, the required personnel are acquired, either through new hires or the transfer of employees, who may be able to bring previous experience to the new product setup. Training is conducted in accordance with process requirements.
10. *Operations Planning*: The development of operational flow charts, raw-material requirements, production rates, maintenance schedules, input/output flows, operating costs, overhead allocation, job routing forms, capacity plans, shift schedules, operations sequences, labor classifications, space requirements, deadlines, and production standards.

11. *Facilities Layout*: Setup of the physical structure or the production line. It may involve the relay out of the existing facility, the installation of new equipment, or the construction of new structures.

12. *System Integration*: Coordinating the new production line to coexist with other lines within the manufacturing system. It may require schedule adjustments to satisfy shared-resource requirements, the development of policies to accommodate product integration, or a realignment of managerial responsibilities.

13. *Production Scheduling*: The generation of schedules of the various activities in the production process. This covers machine assignments, labor assignments, work releases, material supply, in-process storage, and time standards.

14. *Production Run*: The actual implementation of the production schedule. Production control functions may be incorporated into this task. Inspection should be included in the task definitions under "production run" or may be treated as a separate function.

15. *Product Shipment*: This is the final task to complete the project.

The tasks presented in the example may be treated in detail as a subproject of the overall industrial project. In specific situations, some tasks may be added, combined, eliminated, or defined in alternate terms.

INDUSTRIAL PROJECT INTERFACES

The interface between project management functions and the industrial enterprise is easily observable throughout any organization. Many functions that directly or indirectly support the industrial effort can best be managed using project management concepts. Figure 1.11 shows a typical interface of project management in an industrial organization.

Some of the specific tasks and issues to be addressed when managing industrial projects are

- Project selection and prioritizing
- Resource requirements planning
- Cost estimation
- Team formation
- Facility design and management
- Project inventory analysis
- Project forecasting
- Activity modeling
- Human resource management
- Multiproject coordination
- Management of global interfaces
- Project economics
- Contract procurement.

All of these fall within the purview of a formal project management approach.

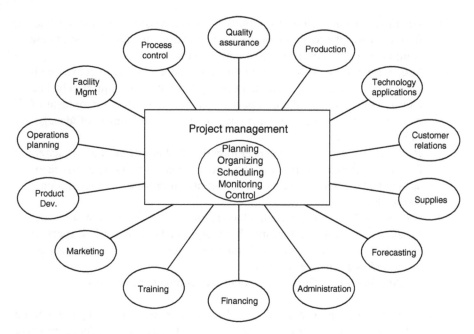

FIGURE 1.11   Project management interfaces in industrial organizations.

PRODUCT QUALITY MANAGEMENT

Quality is a measure of customer satisfaction and a product's "fit-for-use" status. To perform its intended functions, a product must provide a balanced level of satisfaction to both the producer and the customer. For that purpose, we present the following comprehensive definition of quality:

> Quality refers to an equilibrium level of functionality possessed by a product or service based on the producer's capability and the customer's needs.

Based on this definition, quality refers to the combination of characteristics of a product, process, or service that determines the product's ability to satisfy specific needs. Quality is a product's ability to conform to specifications, where specifications represent the customer's needs or government regulations. The attainment of quality in a product is the responsibility of every employee in an organization, and the production and preservation of quality should be a commitment that extends all the way from the producer to the customer. Products that are designed to have high quality cannot maintain the inherent quality at the user's end of the spectrum if they are not used properly.

The functional usage of a product should match the functional specifications for the product within the prevailing usage environment. The ultimate judge for the quality of a product, however, is the perception of the user, and differing circumstances may alter that perception. A product that is perceived as being of high quality for one purpose at a given time may not be seen as having acceptable quality for another purpose in another time frame. Industrial quality standards provide a common basis for global commerce. Customer satisfaction or production efficiency cannot be achieved without product standards. Regulatory, consensus, and contractual requirements should be taken into account when developing product standards. These are described as follows:

Regulatory Standards

This refers to standards that are imposed by a governing body, such as a government agency. All firms within the jurisdiction of the agency are required to comply with the prevailing regulatory standards.

Consensus Standards

This refers to a general and mutual agreement between companies to abide by a set of self-imposed standards.

Contractual Standards

Contractual standards are imposed by the customer based on case-by-case or order-by-order needs. Most international standards will fall into the category of consensus standards, simply because a lack of an international agreement often leads to trade barriers.

INDUSTRIAL PRODUCT DESIGN

The initial step in any manufacturing effort is the development of a manufacturable and marketable product. An analysis of what is required for a design and what is available for the design should be conducted in the planning phase of a design project. The development process must cover analyses of the product configuration, the raw materials required, production costs, and potential profits. Design engineers must select appropriate materials, the product must be expected to operate efficiently for a reasonable length of time (reliability and durability), and it must be possible to manufacture the product at a competitive cost. The design process will be influenced by the required labor skills, production technology, and raw materials. Product planning is substantially influenced by the level of customer sophistication, enhanced technology, and competition pressures. These are all project-related issues that can be enhanced by project management. The designer must recognize changes in all these factors and incorporate them into the design process. Figure 1.12 shows the wide spectrum of what goes into a product design project.

Design in project management provides a guideline for the initiation, implementation, and termination of a design effort. It sets guidelines for specific design objectives, structure, tasks, milestones, personnel, cost, equipment, performance, and problem resolutions. The steps involved include planning, organizing, scheduling, and control. Figure 1.13 shows the various constraints, internal and external, that may affect the design environment.

The availability of technical expertise within an organization and outside of it should be reviewed. The primary question of whether or not a design is needed at

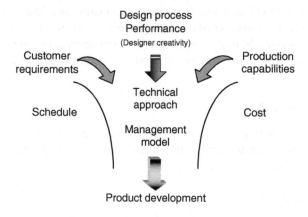

FIGURE 1.12   Product design process.

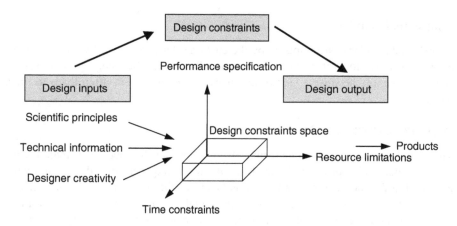

FIGURE 1.13   Industrial design constraints.

all should be addressed. The "make" or "buy," "lease" or "rent," and "do nothing" alternatives to a proposed design should be among the considerations.

In the initial stage of design planning, the internal and external factors that may influence the design should be determined and given relative weights according to priority. Examples of such influential factors include organizational goals, labor situations, market profile, expected return on design investment, technical manpower availability, time constraints, state of the technology, and design liabilities. The desired components of a design plan include summary of the design plan, design objectives, design approach, implementation requirements, design schedule, required resources, available resources, design performance measures, and contingency plans.

Design Feasibility

The feasibility of a proposed design can be ascertained in terms of technical factors, economic factors, or both. A feasibility study is documented with a report showing all the ramifications of the design. A report of the design's feasibility should cover statements about the need, the design process, the cost feasibility, and the design effectiveness. The need for a design may originate from within the organization, from another organization, from the public, or from the customer. Pertinent questions for design feasibility review include: Is the need significant enough to warrant the proposed design? Will the need still exist by the time the design is finished? What are alternate means of satisfying the need? What technical interfaces are required for the design? What is the economic impact of the need? What is the return, financially, on the design change?

A design breakdown structure (DBS) is a flowchart of design tasks required to accomplish design objectives. Tasks that are contained in the DBS collectively describe the overall design. The tasks may involve hardware products, software products, services, and information. The DBS helps to describe the link between the end objective and its components. It shows design elements in the conceptual framework for the purposes of planning and control. The objective of developing a DBS is to study the elemental components of a design project in detail, thus permitting a "divide and conquer" approach. Overall design planning and control can be significantly improved by using DBS. A large design may be decomposed into smaller subdesigns, which may, in turn, be decomposed into task groups. Definable subgoals of a design problem may be used to determine appropriate points at which to decompose the design.

Individual components in a DBS are referred to as *DBS elements* and the hierarchy of each is designated by a level identifier. Elements at the same level of subdivision are said to be of the same DBS level. Descending levels provide increasingly detailed definition of design tasks. The complexity of a design and the degree of control desired are used to determine the number of levels in a DBS. Level I of a DBS contains only the final design purpose. This item should be identifiable directly as an organizational goal. Level II contains the major subsections of the design. These subsections are usually identified by their contiguous location or by their related purpose. Level III contains definable components of the level II subsections. Subsequent levels are constructed in more specific details depending on the level of control desired. If a complete DBS becomes too crowded, separate DBSs may be drawn for the level II components for example. A specification of design (SOD) should accompany the DBS. A statement of design is a narrative of the design to be generated. It should include the objectives of the design, its nature, the resource requirements, and a tentative schedule. Each DBS element is assigned a code (usually numeric) that is used for the element's identification throughout the design life cycle.

Design Stages

The guidelines for the various stages in the life cycle of a design can be summarized in the following way:

1. *Definition of Design Problem*: Define the problem and specify its importance, emphasize the need for a focused design problem, identify designers willing to contribute expertise to the design process, and disseminate the design plan.

2. *Personnel Assignment*: The design group and the respective tasks should be announced and a design manager should be appointed to oversee the design effort.
3. *Design Initiation*: Arrange organizational meeting, discuss general approach to the design problem, announce specific design plan, and arrange for the use of required hardware and tools.
4. *Design Prototype*: Develop a prototype design, test an initial implementation, and learn more about the design problem from test results.
5. *Full-Design Development*: Expand the prototype design and incorporate user requirements.
6. *Design Verification*: Get designers and potential users involved, ensure that the design performs as designed, and modify the design as needed.
7. *Design Validation*: Ensure that the design yields the expected outputs. Validation can address design performance level, deviation from expected outputs, and the effectiveness of the solution to the problem.
8. *Design Integration*: Implement the full design, ensure the design is compatible with existing designs and manufacturing processes, and arrange for design transfer to other processes.
9. *Design Feedback Analysis*: What are the key lessons from the design effort? Were enough resources assigned? Was the design completed on time? Why, or Why not?
10. *Design Maintenance*: Arrange for continuing technical support of the design and update design as new information or technology becomes available.
11. *Design Documentation*: Prepare full documentation of the design and document the administrative process used in generating the design.

INDUSTRIAL OUTSOURCING

Economic pressures on manufacturers have necessitated searching internationally for areas of competitive operations, particularly in terms of labor cost. The availability of specialized skills around the world drives companies to seek skilled personnel at the lowest possible cost from wherever possible. Meanwhile, communication and information technologies have extended the reach and speed of overseas outsourcing, even without sufficient evolution of cultural adaptation. For overseas industrial outsourcing to be successful, both the "source" and "sink" of two cultures must be amenable to cultural integration. Changes to one culture as a result of influences from another culture have historically been very gradual, spanning generations. But in the modern information age, we observe a leap of cultures from one base to another. Cultures of the world evolved along geometrically unique lines over thousands of years. These cultures have been instrumental in defining who, what, how, when, and where of societal endeavors. In recent times, modern information tools and practices have allowed different cultures of the world to interact with one another in ways not previously possible. Education, travel, commerce, industry, and warfare are some of the developments that have facilitated intercultural exposures. Increasing interests in industrial outsourcing to overseas locations have created opportunities for additional cultural interfaces, which can either lead to enmeshing of cultures or creation of conflicts of cultures. This may respectively enhance or impede industrial projects.

Overseas Labor Costs

Some cultures, by their inherent nature, offer lower labor costs for industrial operations. Consequently, the search for a competitive operating site might take a manufacturer to a culture that is totally different from that of the base station. The pervasiveness of fast routing of information makes it easy for a manufacturer to find a seemingly receptive overseas site to relocate operations. The consequence of such relocation is a transfer of culture in either direction. Unfortunately, manufacturers do not fully appreciate the differences in the operating cultures. They pay attention only to the physical and economic aspects of their operations. But more often than not, the cultural shock and unsuccessful assimilation (from either end of the culture transfer) can spell economic failure for the manufacturer.

The economically underdeveloped countries of Asia, Africa, Oceania, and Latin America are among the developing nations. They have common characteristics, such as distorted and highly dependent economies (devoted to producing primary products for the developed world), traditional, rural social structures, high population growth, and widespread poverty. The Third World includes countries on various levels of economic development. Certain characteristics exist that may constitute a barrier to successful implementation of outsourcing to those countries. Some of these are

- Limited access to information (substandard telecommunications infrastructure)
- Political trade barriers
- Cultural norms that impede free flow of information
- Existence and abundance supply of cheap, albeit untrained, workforce
- Orientation of manpower toward artisan and apprenticeship labor

Cultural and Workforce Nuances

Most outsourcing points are located in developing and underdeveloped nations. These locations often have repressive cultures that are replete with norms that the Western world would find unacceptable. A cultural bridge usually is missing between the developed nations and the developing nations with respect to outsourced industrial labor. Thus, there are increasing cultural and economic disparities between global business partners in the selection of outsourcing points. Some of the local issues to be factored into outsourcing projects are summarized below:

- Poverty
- Pollution
- Disease
- Poor hygiene
- Political oppression
- Gender biases
- Economic and financial scams
- Wealth inequities
- Sexual subjugation of women
- Social and sexual permissiveness among elites.

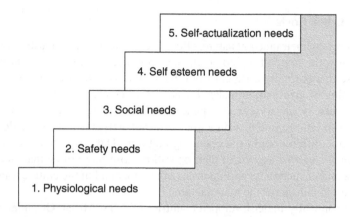

FIGURE 1.14   Typical hierarchy of needs of industrial workers.

Industrial outsourcing representatives posted to these regions are shocked by the level of cultural differences that they observe. In some cases, they maintain a "laissez-faire" and a hands-off attitude. But there have also been cases where some of these representatives take advantage of the loose culturally acceptable social contacts that could be to the detriment of an industrial project.

Figure 1.14 shows the typical hierarchy of needs of workers. In a culturally sensitive workforce, the specific levels and structure of the needs may be drastically different from the typical mode. In an outsource point (sink), most workers will be at the basic level of physiological needs; and there will exist cultural constraints on moving from one level to the next higher level. This fact has an implication on how cultural interfaces can occur between hosts and guests in industrial outsourcing scenarios.

A culture-induced disparity in hierarchy of needs implies that an outsourcing company should recognize how to properly deal with the typical needs of the workforce at each level. The levels are described as follows:

1. *Physiological Needs*: The needs for the basic necessities of life, such as food, water, housing, and clothing (survival needs). This is the level where access to wages is most critical. The basic needs are of primary concern in an outsource location.
2. *Safety Need*: The needs for security, stability, and freedom from threat of physical harm.
3. *Social Needs*: The needs for social approval, friends, love, affection, and association. Industrial outsourcing may bring about better economic outlook that may enable each individual to be in a better position to meet his or her social needs.
4. *Self Esteem Needs*: The needs for accomplishment, respect, recognition, attention, and appreciation. These needs are important not only at the individual level, but also at the organizational level.
5. *Self-Actualization Needs*: These are the needs for self-fulfillment and self-improvement. They also involve the stage of opportunity to grow professionally.

Industrial outsourcing may create opportunities for individuals to assert themselves socially and economically.

Technology and People Factors for Outsourcing

In general, information technology (IT) for outsourcing purposes has definite requirements that may be viewed as qualitative or quantitative and may be divided into technology factors and people factors, as summarized below:

Technology Factors

- Commitment of investment in IT programs
- Hardware installation
- Hardware and software maintenance
- Frequency of enhancements and upgrades
- Reciprocity with other information outlets.

People Factors

- Recruitment of skilled workforce
- Training of the workforce for specific outsourcing needs
- Retention programs for experienced workforce
- Adaptability of the workforce to new modes of business operations.

The level of performance with respect to these factors will affect the feasibility and probability of success of outsourcing implementation. The feasibility of an outsourcing proposal may be determined from several perspectives covering cultural, social, administrative, technical, and economical issues. Typically, the technical and economic aspects get more attention. But in reality, it may be the cultural and social aspects that determine the success of industrial outsourcing.

Cultural and Social Compatibility Issues

Cultural infeasibility is one of the major impediments to outsourcing in an emerging economy. The business climate of today is very volatile. This volatility, coupled with cultural limitations, creates problematic operational elements in a developing country. The pervasiveness of Internet information overwhelms the strict cultural norms in most developing countries. The cultural feasibility of information-based outsourcing needs to be evaluated from the standpoint of where information originates, where it is intended to go, and who comes into contact with the information. For example, the revelation of personal information is viewed as taboo in many developing countries. Consequently, this impedes the collection, storage, and distribution of workforce information that may be vital to the success of outsourcing. For outsourcing to be successfully implemented in such settings, assurances must be incorporated into the hardware and software implementations so as to conciliate the workforce. Accidental or deliberate mismanagement of information is a more worrisome aspect of IT than it is in the Western world, where enhanced techniques are available to correct information errors. What is socially acceptable in the outsourcing culture may not be acceptable in the receiving culture, and vice versa.

Administrative Compatibility

Administrative or managerial feasibility involves the ability to create and sustain an infrastructure to support an operational goal. Should such an infrastructure not be in existence or be unstable, we then have a case of administrative infeasibility. In developing countries, a lack of trained manpower precludes a stable infrastructure for some types of industrial outsourcing. Even where trained individuals are available, the lack of coordination makes it almost impossible to achieve a collective and dependable workforce. Systems that are designed abroad for implementation in a different setting frequently get bogged down when imported into a developing environment that is not conducive for such systems. Differences in the perception of ethics are also an issue of concern in an outsource location. A lack of administrative vision and limited managerial capabilities limit the ability of outsource managers in developing countries. Both the physical and conceptual limitations on technical staff lead to administrative infeasibility that must be reckoned with. Overzealous entrepreneurs are apt to jump on opportunities to outsource production without a proper assessment of the capabilities of the receiving organization. More often than not outsourcing organizations do not fully understand the local limitations. Some organizations take the risk of learning as they go without adequate prior preparation.

Technical Compatibility

Hardware maintenance and software upgrade are, perhaps, the two most noticeable aspects of technical infeasibility of IT in a developing country. A common mistake is to assume that once you install IT and all its initial components, you have the system for life. This is very far from the truth. The lack of proximity to the source of hardware and software enhancement makes this situation particularly distressing in a developing country. The technical capability of the personnel as well as the technical status of the hardware must be assessed in view of the local needs. Doing an overkill on the infusion of IT just for the sake of keeping up is as detrimental as doing nothing at all.

Role of IT in Outsourcing

The use of modern information communication technology (i.e., the Internet) has produced remarkable changes in the ways we communicate and conduct business. We are witnessing the emergence of virtual teams, which consist of geographically dispersed coworkers whose interaction is facilitated through information communication technologies. This is made possible by new and high levels of global connectivity. Significant advances have been seen within the past few years in how people use the latest IT to conduct both official and personal business. In economically developed countries, most enterprises intricately entwine telecommunications in their day-to-day operations and management. With such integration, the Internet constitutes both a threat and a promise. The threats are due to the ways in which electronic communication has significantly altered the way information is captured, stored, manipulated, disseminated, and used. As a result, organizations are vulnerable to potential misuses and abuses—either through accidental incidents or deliberate intents. The promise offered by the Internet relates to the potential to achieve notable enhancements in business processes. The magnitudes of the promises and threats that the Internet presents are particularly pronounced in developing countries, where there is less infrastructure to support it and fewer policies to control its use in business and personal operations.

The Internet is frequently touted not only as an instrument of information handling but also as a medium for wealth creation. The latter is particularly important for developing nations because of the increasing reliance on industrial outsourcing—a key result of globalization. If the Internet can be the vehicle for enhanced international trade, then it will assist countries to become more export oriented and be more competitive internationally. With the aid of the Internet, more and more developing countries are exporting "services" rather than physical products. Manufacturers in industrialized countries can quickly conduct an assessment of potential labor forces available in developing countries to determine outsource targets. Modern IT has facilitated the following types of outsourcing and international trade transactions:

- Transactions for a service, which is completed entirely on the Internet from selection to purchase and delivery.
- Transactions involving "distribution services" in which a product, whether a good or a service, is selected and purchased online but delivered by conventional means.
- Transactions involving telecommunication or routing function of business information.

In order to achieve a successful marriage of IT transactions and existing business functions, a company must understand the respective characteristics of both. Online transactions are characterized by specific functionalities such as online real-time product presentation (electronic catalog), order entry (electronic shopping), product distribution, customer service, product support, and data acquisition. These functionalities create a business environment that has the following properties:

- Fast paced
- Expensive to initiate
- Requires formal training programs.

Not all of these are possible (or feasible) within the existing electronic infrastructure of developing countries. Consequently, there are possible technical and cultural pitfalls such as:

- Error in communication translation
- Error in communication transmission
- Error in communication assimilation.

Workforce Integration Strategies

Any outsourcing enterprise requires adapting from one form of culture to another. The implementation of a new technology to replace an existing (or a nonexistent) technology can be approached through one of several cultural adaptation options. The following are some suggestions:

Parallel Interface: The host culture and the guest culture operate concurrently (side by side) with mutual respect on either side.

Adaptation Interface: This is the case where either the host culture or the guest culture makes conscious effort to adapt to each others' ways. The adaptation often leads to new (but not necessarily enhanced) ways of thinking and acting.

Superimposition Interface: The host culture is replaced (annihilated or relegated) by the guest culture. This implies cultural imposition on local practices and customs. Cultural incompatibility, for the purpose of business goals, is one reason to adopt this type of interface.

Phased Interface: Modules of the guest culture are gradually introduced to the host culture over a period of time.

Segregated Interface: The host and guest cultures are separated both conceptually and geographically. This used to work well in colonial days. But it has become more difficult with modern flexibility of movement and communication facilities.

Pilot Interface: The guest culture is fully implemented on a pilot basis in a selected cultural setting in the host country. If the pilot implementation works with good results, it is then used to leverage further introduction to other localities.

Hybridization of Cultures

The increased interface of cultures through industrial outsourcing is gradually leading to the emergence of hybrid cultures in many developing countries. A hybrid culture derives its influences from diverse factors, where there are differences in how the local population views education, professional loyalty, social alliances, leisure pursuits, and information management. A hybrid culture is, consequently, not fully embraced by either side of the cultural divide. This creates a big challenge to managing outsourcing projects. Table 1.1 summarizes the pros (benefits) and cons (demerits) of cultural enmeshing.

TABLE 1.1
Pros and Cons of Cultural Interfaces in Outsourcing

Benefits	Concerns
• Global awareness for competition	• Subjugation of one culture to the other
• Culturally diverse workforce	• Cultural differences in work ethics
• Access of international labor force	• Loss of cultural identity
• Intercultural peace and harmony	• Bias and suspicion
• Facilitation of social interaction	• Confusion about cultural boundaries
• Closing of trade gaps	• Nonuniform business vocabulary

Workforce Education and Outsourcing

Education should play a significant mitigation role in addressing the cultural challenges facing industrial outsourcing efforts. Formal education introduces the workforce to the cultures of the world. Informal education (or refresher courses) provides operating guidelines for professionals traveling overseas to coordinate outsourcing projects. Suggested educational approaches include:

- Advance briefings
- International exchange programs for students, personnel, managers, and administrators
- Formal lectures and courses
- Training seminars
- Bilateral cultural meetings
- Roundtable discussions involving government, business, industry, and academia
- Hands-on exercises
- Role playing activities
- Open debates designed to iron out and explain unique cultural traits.

Technology Transfer and Outsourcing

Technology transfer for industrial outsourcing purposes can be achieved in various forms. Three technology transfer modes are recommended here to illustrate basic strategies for getting one outsourcing product from one point (technology source) to another point (technology sink). Technology can be transferred in one or a combination of the following strategies:

1. Transfer of complete e-technology products
2. Transfer of technology procedures and guidelines
3. Transfer of technology concepts, theories, and ideas.

Transfer of Complete Technology Products

In this case, a fully developed product is transferred from a source to a target. Very little product development effort is carried out at the receiving point. However, information about the operations of the product is fed back to the source so that necessary product enhancements can be pursued. So, the technology recipient generates product information, which facilitates further improvement at the technology source. This is the easiest mode of technology transfer and the most tempting. Developing nations are particularly prone to this type of transfer. Care must be exercised to ensure that this type of technology transfer does not degenerate into mere "machine transfer."

Transfer of Technology Procedures and Guidelines

In this technology transfer mode, procedures (e.g., blueprints) and guidelines are transferred from a source to a target. The technology blueprints are implemented locally to generate the desired services and products. The use of local raw materials and manpower is encouraged for the local production. Under this mode, the implementation of the transferred technology procedures can generate new operating procedures that can be fed back to enhance the original technology. With this symbiotic arrangement, a loop system is created whereby both the transferring and the receiving organizations derive useful benefits.

Transfer of Technology Concepts, Theories, and Ideas

This strategy involves the transfer of the basic concepts, theories, and ideas supporting a given technology. The transferred elements can then be enhanced, modified, or customized within local constraints to generate new technology products. The local modifications and enhancements have the potential to generate an identical technology, a new related technology, or a new set of technology concepts, theories, and ideas. These derived products may then be transferred back to the original technology source—a kind of reverse outsourcing. Transferred technology must be implemented to work within local limitations. Local innovation, patriotism, dedication, and cultural flexibility to adapt are required to make outsourcing technology transfer successful.

Model for Industrial Technology Transfer

Three technology transfer modes are suggested in Figure 1.15 for a technology transfer linkage between an outsourcing node and an outsource target.

When implementing the suggested technology transfer modes, an organization should carefully examine the following questions:

- Which technology products are most suitable for transfer to a given outsource location, based on the specific attributes of the products and specific factors at the receiving location?

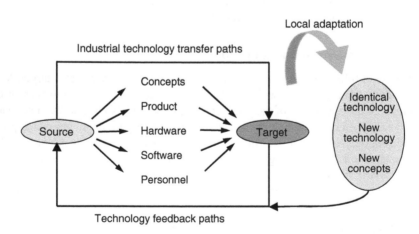

FIGURE 1.15   Industrial technology transfer modes.

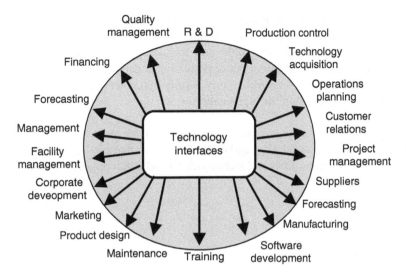

FIGURE 1.16 Outsourcing technology transfer interfaces.

- Which technology procedures and guidelines can be expected to generate the greatest benefits when transferred from a given source to a given target, based on cultural, social, and technical factors at both organizations?
- Which technology concepts, theories, and ideas have the highest potential for practical implementation?

In order to reach the overall goal of a successful technology transfer, it is essential that the most suitable technology be identified, transferred under the most conducive terms, implemented at the receiving organization in the most appropriate manner at the right time, and managed with the utmost commitment. Figure 1.16 shows a model of technology transfer interactions in industrial outsourced operations. Inputs from the environment are capital, labor, and raw material. The synergistic components of the overall system are the production operations, personnel functions, and facilities. These components interact under management guidance to produce products, services, and profitability.

Technology Changeover Strategies

The implementation of a new technology can be effected through one of several strategies. Some strategies are more suitable than others for certain types of outsourcing requirements. The most common changeover strategies include the following:

Parallel Changeover: The existing technology and the new technology operate concurrently until there is confidence that the new technology is satisfactory.
Direct Changeover: The old technology is removed totally and the new technology takes over. This method is recommended only when there is no existing

technology or when both technologies cannot be kept operational because of incompatibility or cost considerations.

Phased Changeover: Modules of the new technology are gradually introduced one at a time using either direct or parallel changeover.

Pilot Changeover: The new technology is fully implemented on a pilot basis in a selected area within the organization.

The specific characteristics of each industrial project determine the level and mode of planning, organizing, scheduling, and control practices that can be utilized. The subsequent chapters present tools and techniques for industrial project management.

2 Principles of Project Management

Project management is the process of managing, allocating, and timing resources in order to achieve a given objective in an expedient manner. The objective may be stated in terms of time (schedule), performance output (quality), or cost (budget). It is the process of achieving objectives by utilizing the combined capabilities of available resources. Time is often the most critical aspect of managing any project. Time is the physical platform over which project accomplishments are made. As emphasized in the poem at the beginning of Chapter 3, time passage of time cannot be stopped. So, it must be managed concurrently with all other important aspects of any project. Project management covers the following basic functions:

1. Planning
2. Organizing
3. Scheduling
4. Controlling.

The complexity of a project can range from simple, such as the painting of a vacant room, to very complex, such as the introduction of a new high-tech product. The technical differences between project types are of great importance when selecting and applying project management techniques. Figure 2.1 illustrates the various dimensions for the application of project management in an industrial system. The analytical approach of activity time modeling presented in Chapter 3 is particularly important for assessing and meeting user requirements in industrial project management.

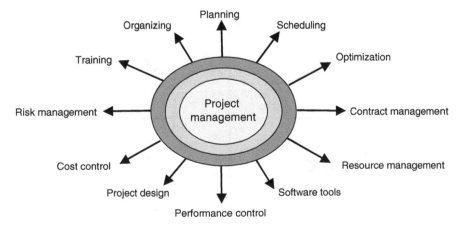

FIGURE 2.1 Dimensions of project management.

Project management techniques are widely used in many human endeavors, such as construction, banking, manufacturing, marketing, health care, sales, transportation, research and development, academics, legal, political, and government establishments, to name just a few. In many situations, the on-time completion of a project is of paramount importance. Delayed or unsuccessful projects not only translate to monetary losses but they also impede subsequent undertakings. Project management takes a hierarchical view of a project environment, covering the top-down levels shown as follows:

1. System level
2. Program level
3. Project level
4. Task level
5. Activity level.

PROJECT REVIEW AND SELECTION CRITERIA

Project selection is an essential first step in focusing the efforts of an organization. Figure 2.2 presents a simple graphical evaluation of project selection. The vertical axis represents the value-added basis of the project under consideration, while the horizontal axis represents the level of complexity associated with the project. In this example, the value can range from low to high, while the complexity can range from easy to difficult. The figure shows four quadrants containing regions of high value with high complexity, low value with high complexity, high value with low complexity, and low value with low complexity. A fuzzy region is identified with an overlay circle. The organization must evaluate each project on the basis of overall organization value streams. The figure can be modified to represent other factors of interest to an organization instead of value-added and project complexity.

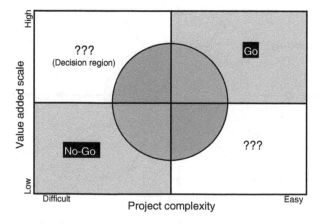

FIGURE 2.2 Project selection quadrant: Go or No-Go.

CRITERIA FOR PROJECT REVIEW

Some of the specific criteria that may be included in project review and selection are as follows:

- Cost reduction
- Customer satisfaction
- Process improvement
- Revenue growth
- Operational responsiveness
- Resource utilization
- Project duration
- Execution complexity
- Cross-functional efficiency
- Partnering potential.

HIERARCHY OF SELECTION

In addition to evaluating an overall project, elements making up the project may need to be evaluated on the basis of the hierarchy presented here. This will facilitate achieving an integrated project management view of the organization's operations.

- System
- Program
- Task
- Work packages
- Activity.

SIZING OF PROJECTS

Associating a size measure to an industrial project provides a means of determining level of relevance and efforts required. A simple guideline is as follows:

- Major (over 60 man-months of effort)
- Intermediate (6–60 man-months)
- Minor (Less than 6 man-months).

PLANNING LEVELS

When selecting projects and their associated work packages, planning should be done in an integrative and hierarchical manner following these levels of planning presented below:

- Supra level
- Macro level
- Micro level.

HAMMERSMITH'S PROJECT ALERT SCALE: RED, YELLOW, GREEN CONVENTION

Hammersmith (2006) presented a guideline for alert scale for project tracking and evaluation. He suggested putting projects into categories of RED, YELLOW, or GREEN with the following definitions:

●   RED (if not corrected, project will be late and/or over budget)
○   YELLOW (project is at risk of turning) RED
◐   GREEN (Project is on time and on budget)

PRODUCT ASSURANCE CONCEPT FOR INDUSTRIAL PROJECTS

Product assurance activities will provide the product deliverables throughout a program development period. These specific activities for continuous effort are to

1. *Track and Incorporate Specific Technologies*: The technology management task will track pertinent technologies through various means (e.g., vendor surveys and literature search). More importantly, the task will determine strategies to incorporate specific technologies.
2. *Analyze Technology Trend and Conduct Long-Range Planning*: The output of technology assessment should be used to formulate long-range policies, directions, and research activities so as to promote the longevity and evolution.
3. *Encourage Government and Industry Leaders' Participation*: In order to determine long-term strategy, the technical evaluation task needs to work closely with government and industry leaders so as to understand their long-range plans. Technology panels may be formed to encourage participation from these leaders.
4. *Influence Industry Directions*: As with other developing programs, the program management will have the opportunity to influence industry direction and spawn new technologies. As the effort can be treated as a model, many technologies and products developed can be applied to other similar systems.
5. *Conduct Prototyping Work*: Prototyping will be used to evaluate the suitability, feasibility, and cost of incorporating a particular technology. In essence, it provides a less costly mechanism to test a technology before significant investment is made in the product development process. Technologies that have high risk with high payoffs should be chosen as the primary subjects for prototyping.

PROJECT MANAGEMENT KNOWLEDGE AREAS

The Project Management Institute (PMI) has identified nine major functional areas that embody the practice of project management. These are compiled into their

Project Management Book of Knowledge (PMBOK). The knowledge areas are summarized here:

1. Scope
2. Cost
3. Schedule
4. Risk
5. Communications
6. Human resources
7. Procurement
8. Quality
9. Integration

The last area covers project integration, health, safety and welfare issues, and professional responsibility. These are topics that are traditionally not addressed explicitly in project management. Several topics covered in this book address many of the key elements embodied in the knowledge areas. Scope management specifies the project charter, project plan, and relationship to other projects going on in an organization. Project integration and scoping are covered in this chapter. Subsequent chapters present topics on time management, resource management, risk management, quality and process management, communications management, and cost management. In spite of the nine distinct areas presented in the PMBOK, all the requirements can be covered in terms of the basic three constraints on any project, as presented earlier in Chapter 1. The triple constraints of time, cost, and performance are shown in terms of the first three elements of PMBOK:

- Scope (Performance) ➔ Performance specs, output targets, and so forth
- Schedule (Time) ➔ Due date expectations, milestones, and so forth
- Cost (Budget) ➔ Budget limitations, cost estimates, and so forth.

If the above three elements are managed effectively, all the other areas will be implicitly or explicitly covered. Cost and schedule are subject to risk. Communications are required for effective scoping. Human resources affect scope, cost, and schedule. Procurement provides the tools and infrastructure for project delivery. Quality implies performance, and vice versa. Integration creates synergy, accountability, and connection among all the elements. Project planning is the basis for achieving adequate attention to all the requirements of a project. The larger and more complex a project, the more critical is the need for using structured project planning.

PROJECT PLANNING

It cannot be overemphasized that planning is the initiation of progress. The key to a successful project is good planning. Project planning provides the basis for the initiation, implementation, and termination of a project, setting guidelines for specific project objectives, project structure, tasks, milestones, personnel, cost, equipment, performance, and problem resolutions. The question of whether or not the project is needed at all, as well as an analysis of what is needed and what is available, should be addressed in the planning phase of new projects. The availability of technical expertise within the organization and outside the organization should be reviewed. If subcontracting is needed, the nature of the contracts should undergo a thorough analysis. The "make," "buy," "lease," "subcontract," or "do-nothing" alternatives should be compared as a part of the project planning process. In the initial stage of project planning, both the internal and external factors that influence the project should be determined and given priority weights. Examples of internal influences on project plans include

- Infrastructure
- Project scope
- Labor relations
- Project location
- Project leadership
- Organizational goal
- Management approach
- Technical manpower supply
- Resource and capital availability.

In addition to internal factors, project plans can be influenced by external factors. An external factor may be the sole instigator of a project, or it may manifest itself in combination with other external and internal factors. Such external factors include the following:

- Public needs
- Market needs
- National goals
- Industry stability
- State of technology
- Industrial competition
- Government regulations.

Strategic planning decisions may be divided into three strategy levels: supra-level planning, macro-level planning, and micro-level planning:

Supra-level Planning: Planning at this level deals with the big picture of how the project fits the overall and long-range organizational goals. Questions faced at this level concern potential contributions of the project to the welfare of the organization, the effect on the depletion of company resources, required interfaces with other projects within and outside the organization, risk exposure, management support for the project, concurrent projects, company culture, market share, shareholder expectations, and financial stability.

Macro-level Planning: Planning decisions at this level address the overall planning within the project boundary. The scope of the project and its operational interfaces should be addressed at this level. Questions faced at the macro level include goal definition, project scope, the availability of qualified personnel and resources, project policies, communication interfaces, budget requirements, goal interactions, deadlines, and conflict resolution strategies.

Micro-level Planning: This deals with detailed operational plans at the task levels of the project. Definite and explicit tactics for accomplishing specific project objectives are developed at the micro level. The concept of management by objective (MBO) may be particularly effective at this level. MBO permits each project member to plan his or her own work at the micro level. Factors to be considered at the micro level of project decisions include scheduled time, training requirements, tools required, task procedures, reporting requirements, and quality requirements.

Large-scale project planning may need to include a statement about the feasibility of subcontracting part of the project work. Subcontracting or outsourcing may be necessary for various reasons, including lower cost, higher efficiency, or logistical convenience.

WORK BREAKDOWN STRUCTURE

Work breakdown structure (WBS) refers to the itemization of a project for planning, scheduling, and control purposes. It presents the inherent components of a project in a structured block diagram or interrelationship flow chart. WBS shows the relative hierarchies of parts (phases, segments, milestone, etc.) of the project. The purpose of constructing a WBS is to analyze the elemental components of the project in detail. If a project is properly designed through the application of WBS at the project planning stage, it becomes easier to estimate cost and time requirements of the project. Project control is also enhanced by the ability to identify how components of the project link together. Tasks that are contained in the WBS collectively describe the overall project goal. Overall project planning and control can be improved by using a WBS approach. A large project may be broken down into smaller subprojects that may, in turn, be systematically broken down into task groups. Thus, WBS permits the implementation of a "divide and conquer" concept for project control.

Individual components in a WBS are referred to as WBS elements, and the hierarchy of each is designated by a level identifier. Elements at the same level of subdivision are said to be of the same WBS level. Descending levels provide increasingly detailed definition of project tasks. The complexity of a project and the degree of control desired determine the number of levels in the WBS. Each component is successively broken down into smaller details at lower levels. The process may continue until specific project activities are reached. In effect, the structure of the WBS looks very much like an organizational chart. The basic approach for preparing a WBS is as follows:

Level 1 WBS: This contains only the final goal of the project. This item should be identifiable directly as an organizational budget item.

Level 2 WBS: This level contains the major subsections of the project. These subsections are usually identified by their contiguous location or by their related purposes.

Level 3 WBS: Level 3 of the WBS structure contains definable components of the level 2 subsections. In technical terms, this may be referred to as the finite element level of the project.

Subsequent levels of WBS are constructed in more specific details depending on the span of control desired. If a complete WBS becomes too crowded, separate WBS layouts may be drawn for the level 2 components. A statement of work (SOW) or WBS summary should accompany the WBS. The SOW is a narrative of the work to be done. It should include the objectives of the work, its scope, resource requirements, tentative due date, feasibility statements, and so on. A good analysis of the WBS structure will make it easier to perform resource work rate analysis using the techniques presented in Chapter 6 of this book.

FEASIBILITY ANALYSIS

The feasibility of a project can be ascertained in terms of technical factors, economic factors, or both. A feasibility study is documented with a report showing all the ramifications of the project and should be broken down into the following categories:

Technical Feasibility: Technical feasibility refers to the ability of the process to take advantage of the current state of the technology in pursuing further improvement. The technical capability of the personnel as well as the capability of the available technology should be considered.

Managerial Feasibility: Managerial feasibility involves the capability of the infrastructure of a process to achieve and sustain process improvement. Management support, employee involvement, and commitment are key elements required to ascertain managerial feasibility.

Economic Feasibility: This involves the ability of the proposed project to generate economic benefits. A benefit–cost analysis and a breakeven analysis are important aspects of evaluating the economic feasibility of new industrial projects. The tangible and intangible aspects of a project should be translated into economic terms to facilitate a consistent basis for evaluation.

Financial Feasibility: Financial feasibility should be distinguished from economic feasibility. Financial feasibility involves the capability of the project organization to raise the appropriate funds needed to implement the proposed project. Project financing can be a major obstacle in large multiparty projects because of the level of capital required. Loan availability, credit worthiness, equity, and loan schedule are important aspects of financial feasibility analysis.

Cultural Feasibility: Cultural feasibility deals with the compatibility of the proposed project with the cultural setup of the project environment. In labor-intensive projects, planned functions must be integrated with the local cultural practices and beliefs. For example, religious beliefs may influence what an individual is willing to do or not do.

Social Feasibility: Social feasibility addresses the influences that a proposed project may have on the social system in the project environment. The ambient social structure may be such that certain categories of workers may be in short supply or nonexistent. The effect of the project on the social status of the project participants must be assessed to ensure compatibility. It should be recognized that workers in certain industries may have certain status symbols within the society.

Safety Feasibility: Safety feasibility is another important aspect that should be considered in project planning. Safety feasibility refers to an analysis of whether the project is capable of being implemented and operated safely with minimal adverse effects on the environment. Unfortunately, environmental impact assessment is often not adequately addressed in complex projects. As an example, the North America Free Trade Agreement (NAFTA) between the US, Canada, and Mexico was temporarily suspended in 1993 because of the legal consideration of the potential environmental impacts of the projects to be undertaken under the agreement.

Political Feasibility: A politically feasible project may be referred to as a "politically correct project." Political considerations often dictate the direction for a proposed project. This is particularly true for large projects with national visibility that may have significant government inputs and political implications. For example, political necessity may be a source of support for a project regardless of the project's merits. On the other hand, worthy projects may face insurmountable opposition simply because of political factors. Political feasibility analysis requires an evaluation of the compatibility of project goals with the prevailing goals of the political system. In general, feasibility analysis for a project should include following items:

1. *Need Analysis*: This indicates recognition of a need for the project. The need may affect the organization itself, another organization, the public, or the government. A preliminary study is conducted to confirm and evaluate the need for the project. A proposal of how the need may be satisfied is then made. Pertinent questions that should be asked include the following:
   - Is the need significant enough to justify the proposed project?
   - Will the need still exist by the time the project is completed?
   - What are alternate means of satisfying the need?
   - What are the economic, social, environmental, and political impacts of the need?

2. *Process Work*: This is the preliminary analysis done to determine what will be required to satisfy the need. The work may be performed by a consultant who is an expert in the project field. The preliminary study often involves system models or prototypes. For technology-oriented projects, artist conceptions and scaled-down models may be used for illustrating the general characteristics of a process. A simulation of the proposed system can be carried out to predict the outcome before the actual project starts.

3. *Engineering and Design*: This involves a detailed technical study of the proposed project. Written quotations are obtained from suppliers and subcontractors as needed. Technology capabilities are evaluated as needed. Product design, if needed, should be done at this stage.

4. *Cost Estimate*: This involves estimating project cost to an acceptable level of accuracy. Levels of around –5% to +15% are common at this level of a project plan. Both the initial and operating costs are included in the cost estimation. Estimates of capital investment, recurring, and non-recurring costs should also be contained in the cost-estimate document. Sensitivity analysis can be carried out on the estimated cost values to see how sensitive the project plan is to changes in the project scenario.

5. *Financial Analysis*: This involves an analysis of the cash-flow profile of the project. The analysis should consider rates of return, inflation, sources of capital, payback periods, breakeven point, residual values, and sensitivity.

6. *Project Impacts*: This portion of the feasibility study provides an assessment of the impact of the proposed project. Environmental, social, cultural, political, and economic impacts may be some of the factors that will determine how a project is perceived by the public. The value-added potential of the project should also be assessed.

7. *Conclusions and Recommendations*: The feasibility study should end with the overall outcome of the project analysis. This may constitute either an endorsement or disapproval of the project.

CONTENTS OF PROJECT PROPOSAL

The project proposal should present a detailed plan for executing the proposed project. The proposal may be directed to a management team within the same organization or to an external organization. The proposal contents may be written in two parts: a technical Section and a Management Section.

TECHNICAL SECTION OF PROJECT PROPOSAL

Project background
- Organization's expertise in the project area
- Project scope
- Primary objectives
- Secondary objectives.

Technical approach
- Required technology
- Available technology
- Problems and their resolutions
- Work breakdown structure.

Work statement
- Task definitions and list
- Expectations.

Schedule
- Gantt charts
- Milestones
- Deadlines.

Project deliverables
The value of the project
- Significance
- Benefit
- Impact.

MANAGEMENT SECTION OF PROJECT PROPOSAL

Project staff and experience
- Personnel credentials.

Organization
- Task assignment
- Project manager, liaison, assistants, consultants, and so forth.

Cost analysis
- Personnel cost
- Equipment and materials
- Computing cost
- Travel
- Documentation preparation
- Cost sharing
- Facilities cost.

Delivery dates
- Specified deliverables.

Quality control measures
- Rework policy.

Progress and performance monitoring
- Productivity measurement.

Cost-control measures

BUDGET PLANNING

The budgeting approach employed for a project can be used to express the overall organizational policy and commitment. Budget often specifies the following:

- Performance measures
- Incentives for efficiency
- Project selection criteria
- Expressions of organizational policy
- Plans for how resources are to be expended
- Catalyst for productivity improvement
- Control basis for managers and administrators
- Standardization of operations within a given horizon.

The preliminary effort in the preparation of a budget is the collection and proper organization of relevant data. However, the preparation of a budget for a project is more difficult than the preparation of budgets for regular and permanent organizational endeavors. While recurring endeavors usually generate historical data that serve as inputs to subsequent estimating functions, projects, on the other hand, are often one-time undertakings without the benefit of prior data. The input data for the budgeting process may include inflationary trends, cost of capital, standard cost guides, past records, and forecast projections. Budgeting may be done as top-down or bottom-up.

TOP-DOWN BUDGETING

This involves collecting data from upper-level sources such as top and middle managers. The cost estimates supplied by the managers may come from their judgments, past experiences, or past data on similar project activities. The cost estimates are passed to lower-level managers, who then break the estimates down into specific work components within the project. These estimates may, in turn, be given to line managers, supervisors, and so on to continue the process. At the end, individual activity costs are developed. The top management presents the overall budget while the line worker generates specific activity budget requirements. One advantage of the top-down budgeting approach is that individual work elements need not be identified prior to approving the overall project budget. Another advantage of the approach is that the aggregate or overall project budget can be reasonably accurate even though specific activity costs may contain substantial errors.

BOTTOM-UP BUDGETING

In bottom-up budgeting, elemental activities, their schedules, descriptions, and labor skill requirements are used to construct detailed budget requests. The line workers who are actually performing the activities are asked to supply cost estimates. Estimates are made for each activity in terms of labor time, materials, and machine time. The estimates are then converted to monetary values. The estimates are combined into composite budgets at each successive level up the budgeting hierarchy. If estimate discrepancies develop, they can be resolved through the intervention to senior management, junior management, functional managers, project managers, accountants, or financial consultants. Analytical tools such as learning-curve analysis, work sampling, and statistical estimation may be used in the budgeting process as appropriate to improve the quality of cost estimates. All component costs and departmental budgets are combined into an overall budget and sent to top management for approval. A common problem with bottom-up budgeting is that individuals tend to overstate their needs with the notion that top management may cut the budget by some percentage. It should be noted, however, that sending erroneous and misleading estimates will only lead to a loss of credibility.

ZERO-BASE BUDGETING

This is another budgeting approach that bases the level of project funding on previous performance. It is normally applicable to recurring programs, especially those in the public sector. Accomplishments in past funding cycles are weighed against the level of resource expenditure. Programs that are stagnant in terms of their accomplishments relative to budget size do not receive additional budgets. Programs that have suffered decreasing yields are subjected to budget cuts or even elimination. By contrast, programs that have a record of accomplishments are rewarded with larger budgets. A major problem with zero-base budgeting is that it puts participants under tremendous pressure to perform data collection, organization, and program justification. So much time may be spent documenting program accomplishments that productivity improvements on current projects may be compromised. Proponents of zero-base budgeting see it as a good approach of encouraging managers and administrators to be more conscious of their management responsibilities. From a project control perspective, the zero-base budgeting approach may be useful in identifying and eliminating specific activities that have not contributed to project goals in the past.

PROJECT ORGANIZATION

Project organization specifies how to integrate the functions of the personnel and stakeholders of a project. Organizing is usually done concurrently with project planning, and directing is an important aspect of organization. Directing involves guiding and supervising the project personnel and is a crucial aspect of the management function, requiring, as it does, skillful managers who can interact with subordinates effectively through good communication and motivation techniques, as well as through proper supervision and delegation of authority. A good project manager can facilitate project success by directing the project team through task assignments, toward the project goal. A good manager recognizes that workers perform better when given clearly defined goals and expectations. A good project manager recognizes individual worker needs and limitations and directs workers' efforts accordingly. Workers should be given some flexibility for self-direction in performing their functions whenever possible so as to increase their sense of investment in and dedication to the project.

THE PROJECT MANAGER

The role of a manager is to use available resources (manpower and tools) to accomplish goals and objectives. A project manager has the primary responsibility of ensuring that a project is implemented according to the project plan. The project manager has a broad sphere of interaction both within and outside the project environment. He or she must be versatile, assertive, and effective in handling problems that develop during the execution phase of the project. Selecting a project manager requires careful consideration and is one of the most crucial aspects of initiating a project. Some aspects of a project require ability to manage while others require ability to lead the project team. Some of the desirable attributes of a project manager are summarized as follows:

- Inquisitiveness
- Good labor relations
- Good motivational skills
- Availability and accessibility
- Versatility with company operations
- Good rapport with senior executives

FIGURE 2.3   Superhero of project management.

- Good analytical and technical background
- Technical and administrative credibility
- Perseverance towards project goals
- Excellent communication skills
- Receptivity to suggestions
- Good leadership qualities
- Good diplomatic skills
- Congenial personality.

The cartoon in Figure 2.3 presents a satirical rendition of the expected versatility and robustness of a project manager.

Just like a Superhero, the project manager must hold the aces in many aspects of project planning, organizing, scheduling, resource allocation, coordination, and control.

PROJECT TEAM MOTIVATION

Motivation is an essential component of implementing project plans. Those who will play direct roles in the project must be motivated to ensure productive participation. Direct beneficiaries of the project must be motivated to make good use of the outputs of the project. Other groups must be motivated to play supporting roles. Motivation may take several forms. For projects that are of a short-term nature, motivation could either be impaired or enhanced by the strategy employed. Impairment may occur if a participant views the project as a mere disruption of regular activities or as a job without long-term benefits. Long-term projects have the advantage of giving participants enough time to readjust to the project efforts.

Theory X Motivation Principle

Theory X assumes that the worker is essentially uninterested and unmotivated to perform his or her work. Motivation must be instilled into the worker by the adoption of external motivating agents. A Theory X worker is inherently indolent and requires constant supervision and prodding to get him to perform. To motivate a Theory X worker, a mixture of managerial actions may be needed. Examples of motivation approaches under Theory X include:

- Rewards to recognize improved effort
- Strict rules to constrain worker behavior
- Incentives to encourage better performance
- Threats to job security associated with performance failure.

Theory Y Motivation Principle

Theory Y assumes that the worker is naturally interested and motivated to perform his or her job. The worker views his or her job function positively and uses self-control and self-direction to pursue project goals. Under Theory Y, management has the task of taking advantage of the worker's readiness and positive intuition so that his or her actions coincide with project objectives. Thus, a Theory Y manager attempts to use the worker's self-direction as the principal instrument for accomplishing work. In general, Theory Y management encourages the following:

- Worker-designed job methodology
- Worker participation in decision making
- Cordial management worker relationship
- Worker individualism within acceptable company limits.

WORKERS HIERARCHY OF NEEDS

The needs of project participants must be taken into consideration in any project planning. Using Maslow's hierarchy of needs, the project leadership must recognize the levels of the basic needs of work groups as summarized as follows:

1. *Physiological Needs*: The most basic needs of life: food, water, housing, and clothing. This is the level where access to money is most critical.
2. *Safety Needs*: The need for security, stability, and freedom from threat of physical harm. The fear of adverse environmental impact may inhibit project efforts.
3. *Social Needs*: The need for social approval, friends, love, affection, and association. For example, public service projects may bring about a better economic situation, which may in turn enable individuals to be in a better position to meet their social needs.

4. *Esteem Needs*: the needs for accomplishment, respect, recognition, attention, and appreciation. These needs are important not only at the individual level, but also at the community level, extending even to the national level.
5. *Self-Actualization Needs*: The needs for self-fulfillment and self-improvement. They also involve the availability of opportunity to grow professionally. Work improvement projects may lead to self-actualization opportunities for individuals to assert themselves socially and economically. Job achievement and professional recognition are two of the most important factors that lead to employee satisfaction and better motivation.

Hierarchical motivation implies that the particular motivation technique utilized at a given level may not be applicable at a different level.

MOTIVATING AND DEMOTIVATING FACTORS

Herzberg's motivation concept takes a look at the characteristics of work itself as the motivating factor. There are two motivational factors, classified as the hygiene factors and motivators, which state that the hygiene factors are necessary but not sufficient conditions for a contented worker. The negative aspects of the factors may lead to a disgruntled worker, whereas the positive aspects do not necessarily enhance motivation. Examples include:

1. *Administrative Policies*: Bad policies can lead to the discontent of workers, whereas good policies are viewed as routine with no specific contribution to improving worker satisfaction.
2. *Supervision*: A bad supervisor can make a worker unhappy and less productive, but a good supervisor cannot necessarily improve worker performance.
3. *Working Condition*: Bad working conditions can enrage workers, but good working conditions do not automatically generate improved productivity.
4. *Salary*: Low salaries can make a worker unhappy, disruptive, and uncooperative, but a raise will not necessarily provoke him to perform better. Although a raise in salary will not necessarily increase professionalism, a reduction in salary will most certainly have an adverse effect on morale.
5. *Personal Life*: Miserable personal life can adversely affect worker performance, but a happy life does not imply that he will be a better worker.
6. *Interpersonal Relationships*: Good peer, superior, and subordinate relationships are important to keep a worker happy and productive, but extraordinarily good relations do not guarantee that he will be more productive.
7. *Social and Professional Status*: Low status can force a worker to perform at his low "level," whereas high status does not imply that he will perform at a higher level.
8. *Security*: A safe environment may not motivate a worker to perform better, but unsafe conditions will certainly impede his productivity.

Motivators are motivating agents that should be inherent in the work itself. If necessary, project task assignments should be redesigned (or reengineered) to include inherent motivating factors. Some guidelines:

1. *Achievement*: The job design should facilitate opportunity for worker achievement and advancement toward personal goals.
2. *Recognition*: The mechanism for recognizing superior performance should be incorporated into the task assignment. Opportunities for recognizing innovation should be built into the task.
3. *Work Content*: The work content should be interesting enough to motivate and stimulate the creativity of the worker. The amount of work and the organization of the work should be designed to fit a worker's needs.
4. *Responsibility*: The worker should have some measure of responsibility for how his or her job is performed. Personal responsibility leads to accountability which leads to better performance.
5. *Professional Growth*: The work should offer an opportunity for advancement so that the worker can set his own achievement level for professional growth within a project plan.

MANAGEMENT BY OBJECTIVE

Management by objective (MBO) is the management concept whereby a worker is allowed to take responsibility for the design and performance of a task under controlled conditions. It gives each worker a chance to set his or her own objectives in achieving project goals. The worker can monitor his own progress and take corrective actions when needed without management intervention. Workers under the concept of Theory Y appear to be the best suited for the MBO concept. However, MBO has some disadvantages, which include the possible abuse of the freedom to self-direct and the possible disruption of overall project coordination. The advantages of MBO include the following:

1. Encouraging each worker to find better ways of performing the job
2. Avoiding the oversupervision of professionals
3. Helping workers become better aware of what is expected of them
4. Permitting timely feedback on worker performance.

MANAGEMENT BY EXCEPTION

Management by exception (MBE) is an after-the-fact management approach to the issue of control. Contingency plans are not made and there is no rigid monitoring. Deviations from expectations are viewed as exceptions to the normal courses of events. When intolerable deviations from plans occur, they are investigated, and then an action is taken. The major advantage of MBE is that it lessens the management workload and reduces the cost of management. However, it is a risky concept to follow, especially for high-stake industry projects. Many of the problems that can develop in complex projects are such that after-the-fact corrections are expensive or impossible.

PROJECT ORGANIZATION STRUCTURES

After project planning, the next step is the selection of the project organizational structure. Before selecting an organizational structure, the project team should assess the nature of the job to be performed and its requirements. The structure may be defined in terms of functional specializations, departmental proximity, standard management boundaries, operational relationships, or product requirements.

TRADITIONAL FORMAL ORGANIZATION STRUCTURES

Many organizations use the traditional formal or classical organization structures, which show hierarchical relationships between individuals or teams of individuals. Traditional formal organizational structures are effective in service enterprises because groups with similar functional responsibilities are clustered at the same level of the structure. A formal organizational structure represents the officially sanctioned structure of a functional area. An informal organizational structure, on the other hand, develops when people organize themselves in an unofficial way to accomplish a project objective. The informal organization is often very subtle in that not everyone in the organization is aware of its existence. Both formal and informal organizations exist within every project. Positive characteristics of the traditional formal organizational structure include the following:

- Availability of broad manpower base
- Identifiable technical line of control
- Grouping of specialists to share technical knowledge
- Collective line of responsibility
- Possibility of assigning personnel to several different projects
- Clear hierarchy for supervision
- Continuity and consistency of functional disciplines
- Possibility for the establishment of departmental policies, procedures, and missions.

However, the traditional formal structure does have some negative characteristics as summarized here:

- No one individual is directly responsible for the total project
- Project-oriented planning may be impeded
- There may not be a clear line of reporting up from the lower levels
- Coordination is complex
- A higher level of cooperation is required between adjacent levels
- The strongest functional group may wrongfully claim project authority.

FUNCTIONAL ORGANIZATION

The most common type of formal organization is known as the functional organization, whereby people are organized into groups dedicated to particular functions. Depending on the size and the type of auxiliary activities involved, several minor, but supporting, functional units can be developed for a project. Projects that are organized along functional lines normally reside in a specific department or area of specialization. The project home office or headquarters is located in the specific functional department.

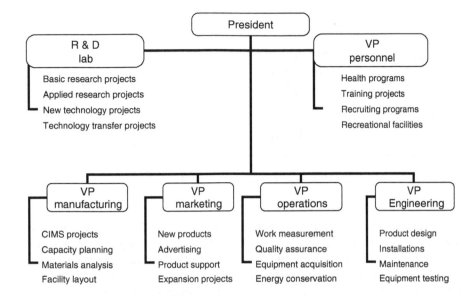

FIGURE 2.4 Functional organization structure.

Figure 2.4 shows examples of projects that are functionally organized. The advantages of a functional organization structure are:

- Improved accountability
- Discernible lines of control
- Flexibility in manpower utilization
- Enhanced comradeship of technical staff
- Improved productivity of specially skilled personnel
- Potential for staff advancement along functional path
- Ability of the home office to serve as a refuge for project problems.

The disadvantages of a functional organization structure include:

- Potential division of attention between project goals and regular functions
- Conflict between project objectives and regular functions
- Poor coordination similar project responsibilities
- Unreceptive attitudes on the part of the surrogate department
- Multiple layers of management
- Lack of concentrated effort.

PRODUCT ORGANIZATION

Another approach to organizing a project is to use the end product or goal of the project as the determining factor for personnel structure. This is often referred to as pure project organization or simply project organization. The project is set up as a

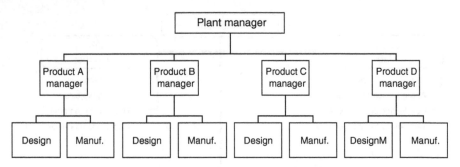

FIGURE 2.5   Product organization structure.

unique entity within the parent organization. It has its own dedicated technical staff and administration. It is linked to the rest of the system through progress reports, organizational policies, procedures, and funding. The interface between product-organized projects and other elements of the organization may be strict or liberal, depending on the organization. An example of a pure project organization is shown in Figure 2.5.

The product organization is common in industries that have multiple product lines. Unlike the functional, the product organization decentralizes functions. It creates a unit consisting of specialized skills around a given project or product. Sometimes referred to as a team, task force, or product group, the product organization is common in public, research, and manufacturing organizations where specially organized and designated groups are assigned specific functions. A major advantage of the product organization is that it gives the project members a feeling of dedication to and identification with a particular goal.

A possible shortcoming of the product organization is the requirement that the product group be sufficiently funded to be able to stand alone. The product group may be viewed as an ad hoc unit that is formed for the purpose of a specific goal. The personnel involved in the project are dedicated to the particular mission at hand. At the conclusion of the mission, they may be reassigned to other projects. Product organization can facilitate the most diverse and flexible grouping of project participants. It has the following advantages:

- Simplicity of structure
- Unity of project purpose
- Localization of project failures
- Condensed and focused communication lines
- Full authority of the project manager
- Quicker decisions due to centralized authority
- Skill development due to project specialization
- Improved motivation, commitment, and concentration
- Flexibility in determining time, cost, performance trade-offs
- Project team's reporting directly to one project manager or boss
- Ability of individuals to acquire and maintain expertise on a given project.

The disadvantages of product organization are

- Narrow view on the part of project personnel (as opposed to a global organizational view)
- Mutually exclusive allocation of resources (one worker to one project)
- Duplication of efforts on different but similar projects
- Monopoly of organizational resources
- Worker concern about life after the project
- Reduced skill diversification.

One other disadvantage of the product organization is the difficulty supervisors have in assessing the technical competence of individual team members. As managers are leading people in fields foreign to them, it is difficult for them to assess technical capability. Many major organizations have this problem. Those who can talk a good game and give good presentations are often viewed by management as knowledgeable, regardless of their true technical capabilities.

MATRIX ORGANIZATION STRUCTURE

The matrix organization is a frequently used organization structure in industry. It is used where there is multiple managerial accountability and responsibility for a project. It combines the advantages of the traditional structure and the product organization structure. The hybrid configuration of the matrix structure facilitates maximum resource utilization and increased performance within time, cost, and performance constraints. There are usually two chains of command involving both horizontal and vertical reporting lines. The horizontal line deals with the functional line of responsibility while the vertical line deals with the project line of responsibility. An example of a matrix structure is shown in Figure 2.6.

Advantages of matrix organization include the following:

- Good team interaction
- Consolidation of objectives
- Multilateral flow of information
- Lateral mobility for job advancement
- Individuals have an opportunity to work on a variety of projects
- Efficient sharing and utilization of resources
- Reduced project cost due to sharing of personnel
- Continuity of functions after project completion
- Stimulating interactions with other functional teams
- Functional lines rally to support the project efforts
- Each person has a "home" office after project completion
- Company knowledge base is equally available to all projects.

Some of the disadvantages of matrix organization are summarized as follows:

- Matrix response time may be slow for fast-paced projects
- Each project organization operates independently

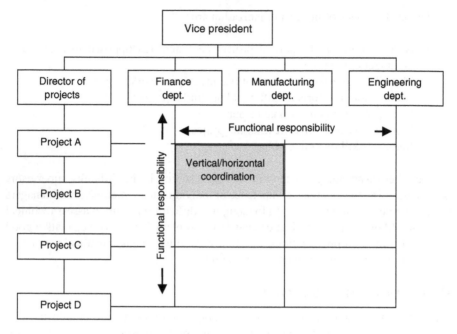

FIGURE 2.6   Matrix organization structure.

- Overhead cost due to additional lines of command
- Potential conflict of project priorities
- Problems inherent in having multiple bosses
- Complexity of the structure.

Traditionally, industrial projects are conducted in serial functional implementations such as research and development (R&D), engineering, manufacturing, and marketing. At each stage, unique specifications and work patterns may be used without consulting the preceding and succeeding phases. The consequence is that the end product may not possess the original intended characteristics. For example, the first project in the series might involve the production of one component while the subsequent projects might involve the production of other components. The composite product may not achieve the desired performance because the components were not designed and produced from a unified point of view. The major appeal of matrix organization is that it attempts to provide synergy within groups in an organization.

PROJECT TEAM BUILDING

Team building is a key part of project management because workers are expected to perform not just on one team but on a multitude of interfacing teams. To facilitate effective team building, it is important to distinguish between functional organization structure and operational organization structure. Within a team, the functional structure is developed according to functional lines of responsibility while the operational

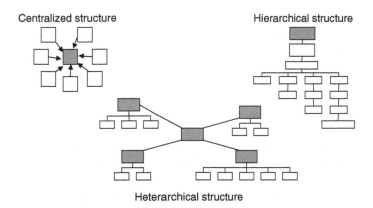

Centralized structure   Hierarchical structure

Heterarchical structure

FIGURE 2.7   Team operational organization structures.

structure takes into account the way workers actually organize themselves to accomplish a task. Figure 2.7 illustrates three common operational organization options within a team structure. The operational organization can be centralized, hierarchical, or heterarchical.

In a centralized structure, workers rely on a specific individual or office for reporting and directive purposes. In a hierarchical structure, the lines of functional reporting existing in the organization are observed as workers pursue their tasks. This approach may discourage free and informal interaction, thereby creating obstacles to effective work. In a heterarchical structure, different heterogeneous teams of workers interact to solve a problem with the aid of a central coordination agent. The effectiveness of a team can be affected by its size and its purpose. Small teams are useful for responsiveness and prompt action. Large teams are useful for achieving a widespread information base, inclusion, and participation. Strategies for enhancing project team building include:

1. Interoffice employee exchange programs
2. Interproject transferability of personnel
3. In-house training for new employees
4. Diversification and development of in-house job skills
5. Use of in-house personnel as in-house consultants.

One of the most important ingredients for establishing a good team is commitment. The Triple C approach can help achieve such commitment. The organizational structure must create an environment that promotes, or even demands, teamwork. The project manager should strive to provide and maintain an atmosphere that fulfills the needs and expectations of workers. A team should offer opportunities for positive professional interactions as well as avenues for advancement. Some of the expectations and opportunities that a team environment should offer are:

• Good team leadership
• Challenging (not problematic) work environment

- Recognition for team accomplishment
- Recognition for individual contributions
- Ability to speak in a collective voice to management
- Opportunities for career growth
- Opportunity for social interaction
- Possibility of both exercising individual discretion and arriving at collective consensus.

Some of the barriers to team building are changes in the work environment, lack of member commitment, poor intrateam communication, lack of management support, team superstar model. On some teams, only one or a few individuals lead the efforts and get all the credit at the expense of the other team members. A teaming approach that allows everyone to pitch in and share credit equitably is preferred. If the team excels, everyone excels. Shared project glory is to everyone's glory.

PROJECT PARTNERING

Project partnering involves having project team and stakeholders operating as partners in the pursuit of project goals. Benefits of project partnering include improved efficiency, cost reduction, resource sharing, better effectiveness, increased potential for innovation, and improvement of quality of products and services. Partnering creates a collective feeling of being together on the project. It fosters a positive attitude that makes it possible to appreciate and accept the views of others. It also recognizes the objectives of all parties and promotes synergism. The communication, cooperation, and coordination concepts of the Triple C model facilitate partnering. Suggestions for setting up project partnering are

- Use an inclusive organization structure
- Identify project stakeholders and clients
- Create informational linkages
- Identify the lead partner
- Collate the objectives of partners
- Use a responsibility chart to assign specific functions.

TEAM DECISION MAKING

For decisions to be effective, they have to be good and timely. Teams often need to make decisions on a group basis. Many decision situations, however, are complex. No one person has all the information to make all decisions accurately. As a result, crucial decisions must often be made by a group of people. The major advantages of group decision making include the following:

1. *Ability to Share Experience, Knowledge, and Resources*: Many heads are better than one. A group will possess greater collective ability to solve a given decision problem.

2. *Increased Credibility*: Decisions made by a group of people often carry more weight in an organization.
3. *Improved Morale*: Staff morale can be positively influenced because many people have the opportunity to participate in the decision-making process.
4. *Better Rationalization*: The opportunity to observe other people's views can lead to an improvement in an individual's reasoning process.

There are several group decision-making approaches, each with pros and cons. They are summarized in the following sections.

PROJECT BRAINSTORMING

Brainstorming is a way of generating many new ideas. In brainstorming, the decision group comes together to discuss alternate ways of solving a problem. The members of the brainstorming group may be from different departments, may have different backgrounds and training, and may not even know one another. The diversity of the participants helps to create a stimulating environment for generating different ideas from different views. The technique encourages free outward expression of new ideas, no matter how far-fetched the ideas might appear. No criticism of any new idea is permitted during the brainstorming session. A major concern in brainstorming is that extroverts may take control of the discussions. For this reason, an experienced and respected individual should manage the brainstorming discussions. The group leader establishes the procedure for proposing ideas, keeps the discussions in line with the group's mission, discourages disruptive statements, and encourages the participation of all members. After the group runs out of more ideas, open discussions are held to weed out the unsuitable ones. It is to be expected that even the rejected ideas may stimulate the generation of other ideas that may eventually lead to other favored ideas. Guidelines for improving brainstorming:

* Focus on a specific decision problem
* Keep ideas relevant to the intended decision
* Be receptive to all new ideas
* Evaluate the ideas on a relative basis after exhausting new ideas
* Maintain an atmosphere conducive to cooperative discussions
* Maintain a record of the ideas generated.

DELPHI METHOD

The traditional approach to group decision making is to obtain the opinion of experienced participants through open discussions. The goal is to reach a consensus among the participants. However, open group discussions are often biased because of the influence of or subtle intimidation from dominant individuals. Even when the threat of a dominant individual is not present, opinions may still be swayed by group pressure. This is called the "bandwagon effect" of group decision making. The Delphi method attempts to overcome these difficulties by requiring individuals to present their opinions anonymously through an intermediary. The method differs from other interactive group methods in that it eliminates face-to-face confrontations. It was originally developed for forecasting applications. But it has been modified in various ways for application to different types of group decision making. The steps of the Delphi method are

1. *Problem Definition*: A decision problem that is considered significant is identified and clearly described.
2. *Group Selection*: An appropriate group of experts or experienced individuals is formed to address the particular decision problem. Both internal and external experts may be involved in the Delphi process. A leading individual is appointed to serve as the administrator of the decision process. The group may operate through mail, e-mail, or it may gather together in a room. In either case, all opinions are expressed anonymously on paper. If the group meets in the same room, care should be taken to provide enough room so that each member does not have the feeling that someone may accidentally or deliberately observe his or her responses.
3. *Initial Opinion Poll*: The technique is initiated by describing the problem to be addressed in unambiguous terms. The group members are requested to submit a list of major areas of concern in their specialty areas as they relate to the decision problem.
4. *Questionnaire Design and Distribution*: Questionnaires are prepared to address the areas of concern related to the decision problem. The written responses to the questionnaires are collected and organized by the administrator. The administrator aggregates the responses in a statistical format. For example, the average, mode, and median of the responses may be computed. This analysis is distributed to the decision group. Each member can then see how his or her responses compare with the anonymous views of the other members.
5. *Iterative Balloting*: Additional questionnaires based on the previous responses are passed to the members. The members submit their responses again. They may choose to alter or not alter their previous responses.
6. *Silent Discussions and Consensus*: The iterative balloting may involve anonymous written discussions of why some responses are correct or incorrect. The process is continued until a consensus is reached. A consensus may be declared after five or six iterations of the balloting or when a specified percentage (e.g., 80%) of the group agrees on the questionnaires. If a consensus cannot be declared on a particular point, it may be displayed to the whole group with a note that it does not represent a consensus.

The Delphi Method's major characteristics of anonymity of responses, statistical summary of responses, and controlled procedure make it a reliable mechanism for obtaining numeric data from subjective opinion. The major limitations of the Delphi method are

1. Its effectiveness may be limited in cultures where strict hierarchy, seniority, and age influence decision-making processes.

2. Some experts may not readily accept the contribution of nonexperts to the group decision-making process.
3. As opinions are expressed anonymously, some members may take the liberty to make ludicrous statements. However, if the group composition is carefully reviewed, this problem may be avoided.

NOMINAL GROUP TECHNIQUE

Nominal group technique is a silent version of the brainstorming method and is another way of reaching consensus. Rather than asking people to state their ideas aloud, the team leader asks each member to jot down a minimum number of ideas, for example, five or six. A single list of ideas is then composed on a chalkboard for the whole group to see. The group then discusses the ideas and weeds out some in an iterative process until a final decision is made. The nominal group technique is relatively easy to control. Unlike brainstorming, where members may get into shouting matches, the nominal group technique permits members to silently present their views. In addition, it allows introverted or shy members to contribute to the decision without the pressure of having to speak out too often. The steps for nominal group technique are summarized as follows:

1. Silently generate ideas, in writing
2. Record ideas without discussion
3. Conduct group discussion for clarification of meaning, not argument
4. Vote to establish the priority or rank of each item
5. Discuss vote
6. Cast final vote.

In all of the group decision-making techniques, an important action that can enhance and expedite the decision-making process is to require that members review all pertinent data before coming to the group meeting. This will ensure that the decision process is not impeded by trivial preliminary discussions

INFORMATION GATHERING TECHNIQUES

Interviews, surveys, and questionnaires are important information-gathering techniques. They also foster cooperative working relationships and encourage direct participation and input into project decision-making processes. They provide an opportunity for employees at the subordinate levels of an organization to contribute ideas and inputs for decision making. The greater the number of people involved in the interviews, surveys, and questionnaires, the more valid will the final decision be. The following guidelines are useful for conducting interviews, surveys, and questionnaires to collect data and information for project decisions.

1. Collect and organize background information and supporting documents on the items to be covered by the interview, survey, or questionnaire.
2. Outline the items to be covered and list the major questions to be asked.
3. Use a suitable medium of interaction and communication, such as telephone, fax, e-mail, face-to-face, observation, meeting venue, poster, or memo.

4. Tell the respondents the purpose of the interview, survey, or questionnaire and indicate how long it would take.
5. Use open-ended questions that stimulate ideas from the respondents.
6. Minimize the use of "yes" or "no" type of questions.
7. Encourage expressive statements that indicate the respondent's views.
8. Use the "who, what, when, where, why, and how" approach to elicit specific information.
9. Thank the respondents for their participation.
10. Let the respondents know the outcome of the exercise.

Multivote

Multivoting is a series of votes used to arrive at a group decision. It can be used to assign priorities to a list of items, especially at team meetings after a brainstorming session has generated a long list of items. Multivoting helps to reduce such long lists to a few items, usually three to five. The steps for multivoting are:

1. Take a first vote. Each person votes as many times as desired, but only once per item.
2. Circle the items receiving a relatively higher number of votes (i.e., majority vote) than the other items.
3. Take a second vote. Each person votes for a number of items equal to one-half the total number of items circled in step 2. Only one vote per item is permitted.
4. Repeat steps 2 and 3 until the list is reduced to three to five items depending on the needs of the group. It is not recommended to multivote down to only one item.
5. Perform further analysis of the items selected in step 4, if needed.

PROJECT TEAM LEADERSHIP

Good leaders lead by example. Others attempt to lead by dictating. It is an element of human nature that people learn and act best when good examples are available for them to emulate, and they generally do not forget the lessons they learn in this way. Good examples observed in childhood, for instance, can provide a lifetime's worth of guidance. A leader should have a spirit of performance that stimulates his or her subordinates to perform at their own best. Rather than dictating what needs to be done, a good leader should show what needs to be done. Showing, in this case, does not necessarily imply actual physical demonstration of what is to be done. Rather, it implies projecting a commitment to the function at hand and a readiness to participate as appropriate. In the traditional model, managers manage workers to get them to work. There may be no point of convergence or active participation. Modern managers, however, team up with workers to get the job done. A good leadership model will encompass listening and asking questions, specifying objectives, developing clear directions, removing obstacles, encouraging individual initiatives, learning from past experiences, reiterating the project requirements, and getting everyone involved. Good leadership lessons can be learned a Chinese proverb that says,

Tell me, and I forget;
Show me, and I remember;
Involve me, and I understand.

Telling, showing, and involving are good practices for project team leadership. Project team leadership involves dealing with managers and supporting personnel across the functional lines of the project. It is a misconception to think that a leader leads only his or her own subordinates. Leadership responsibilities can cover functions vertically up or down. A good project leader can lead and inspire not only his or her subordinates, but also the entire project organization, including the highest superiors. Generally, leadership involves recognizing an opportunity to make improvements in a project and taking the initiative to lead the implementation of the improvements. In addition to inherent personal qualities, leadership style can be influenced by training, experience, and dedication. Guidelines for project team leadership are

- Avoid organizational politics and personal egotism
- Lead by example
- Place principles above personality
- Focus on the big picture of project goals
- Build up credibility with successful leadership actions
- Demonstrate integrity and ethics in decision processes
- Cultivate and encourage a spirit of multilateral cooperation
- Preach less and implement more
- Back up words with action.

PROJECT SCHEDULING

Project scheduling is the time-phased arrangement of project activities subject to precedence, time, and resource constraints in order to accomplish project objectives. Project scheduling is distinguished from industrial job shop, flow shop, and other production sequencing problems because of the nonrepetitive nature of most projects. In production scheduling, the scheduling problem follows a standard procedure that determines the characteristics of production operations. A scheduling technique that works for one production run may be expected to work equally effectively for succeeding and identical production runs. In other words, reliable precedents can be found for production scheduling problems. On the other hand, projects are usually one-time endeavors that are rarely duplicated in identical circumstances. In some cases, it may be possible to duplicate the concepts of the whole project or a portion of it. The construction of a dam is a good example. The concepts of dam construction will, most likely, be the same in all dam operations. It would, however, be highly unlikely to have two dam construction projects that were exactly alike. Even if two projects are identical in many details, manpower availability (a critical component of any project) is likely to be different. This, of course, makes project scheduling more challenging than production scheduling.

Project scheduling represents the core of project management efforts because it involves the assignment of time periods to specific tasks within the work schedule. Resource availability, time limitations, urgency and priority, performance specification, precedence constraints, milestones, technical precedence constraints, and other factors complicate the scheduling process. Generally, project scheduling involves

- Analysis of resource availability in terms of human resources, materials, capital, and so forth
- Application of scheduling techniques (CPM, PERT, Gantt charts, PDM)
- Tracking and reporting
- Control and termination.

PROJECT CONTROL

Project control requires that appropriate actions be taken to correct deviations from expected performance. Control involves measurement, evaluation, and correction. Measurement is the process of measuring the relationship between planned performance and actual performance with respect to project objectives. The variables to be measured, the measurement scales, and the measuring approaches should be clearly specified during the planning stage. Corrective actions may involve rescheduling, reallocation of resources, or expediting of tasks. In some cases, project termination is an element of project control.

THE TRIPLE C MODEL

The Triple C model introduced by Badiru (1987) is an effective tool for project planning and control. The model can facilitate better resource management by identifying the crucial aspects of a project. The model states that project management can be enhanced by its implementation within the integrated functions of:

- Communication
- Cooperation
- Coordination.

The model facilitates a systematic approach to project planning, organizing, scheduling, and control. It highlights what must be done and when. It also helps to identify the resources (manpower, equipment, facilities, etc.) required for each effort. Triple C points out important questions, such as

- Does each project participant know what the objective is?
- Does each participant know his or her role in achieving the objective?
- What obstacles may prevent a participant from playing his or her role effectively?

Figure 2.8 shows a graphical representation of the Triple C model for project management. If resource management is viewed as a three-legged stool, then communication, cooperation, and coordination constitute the three legs. Communication

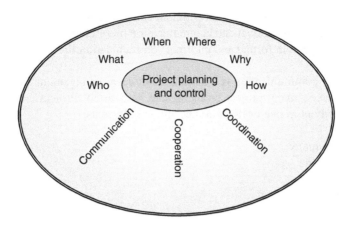

FIGURE 2.8   Triple C model for project management.

channels provide the basis for effective communication, partnership forms the basis for cooperation, and organizational structure provides the basis for coordination. Consequently, there must be appropriate communication channels, partnership, and proper organization structure for Triple C to be effective. This is summarized as follows:

1. For effective communication, create good communication channels.
2. For enduring cooperation, establish partnership arrangements.
3. For steady coordination, use a workable organization structure.

PROJECT COMMUNICATION

The communication function of project management involves making sure that all those concerned become aware of project requirements and progress. Those that will be affected by the project directly or indirectly, as direct participants or as beneficiaries, should be informed as appropriate regarding the following:

- The scope of the project
- The personnel contribution required
- The expected cost and merits of the project
- The project organization and implementation plan
- The potential adverse effects if the project should fail
- The alternatives, if any, for achieving the project goal
- The potential benefits (direct and indirect) of the project.

Communication channels must be kept open throughout the project life cycle. In addition to internal communication, appropriate external sources should also be consulted. The project manager has several "must do" requirements:

- Exude commitment to the project
- Utilize a communication responsibility matrix
- Facilitate multichannel communication interfaces

- Identify internal and external communication needs
- Resolve organizational and communication hierarchies
- Encourage both formal and informal communication links.

When clear communication is maintained between management and employees and among peers, many project problems can be averted. Project communication may be carried out in one or more of the following formats:

- One-to-many
- One-to-one
- Many-to-one
- Written and formal
- Written and informal
- Oral and formal
- Oral and informal
- Nonverbal gesture.

Figure 2.9 shows an example of a design of a communication responsibility matrix. A communication responsibility matrix shows the linking of sources of communication and targets of communication. Cells within the matrix indicate the subject of the desired communication. There should be at least one filled cell in each row and each column of the matrix. This assures that each individual within a department has at least one communication source or target associated with him or her. With a communication responsibility matrix, a clear understanding of what needs to be communicated to whom can be developed.

PROJECT COOPERATION

The cooperation of the project personnel must be explicitly elicited. Merely voicing consent for a project is not enough assurance of full cooperation. The participants

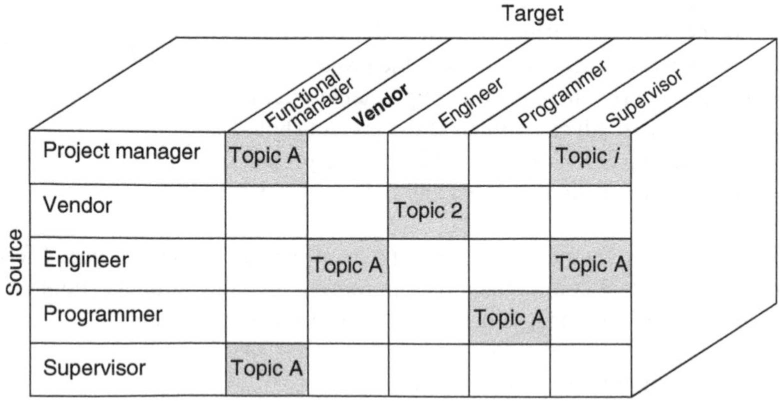

FIGURE 2.9   Triple C communication matrix.

and beneficiaries of the project must be convinced of the merits of the project. Some of the factors that influence cooperation in a project environment include manpower requirements, resource requirements, budget limitations, past experiences, conflicting priorities, and lack of uniform organizational support. A structured approach to seeking cooperation should clarify the following:

- The cooperative efforts required
- The precedents for future projects
- The implications of what lack of cooperation can do to a project
- The criticality of cooperation to project success
- The organizational impact of cooperation .
- The time frame involved in the project
- The rewards for good cooperation.

Cooperation is a basic virtue of human interaction. More projects fail because of lack of cooperation and commitment than any other project factors. To secure and retain the cooperation of project participants, their first reaction to the project must be positive. The most positive aspects of a project should be the first items of project communication. Whichever type of cooperation is available in a project environment, the cooperative forces should be channeled toward achieving project goals. A documentation of prevailing level of cooperation is useful for winning further support for a project. Clarification of project priorities will facilitate personnel cooperation. Relative priorities of multiple projects should be specified so that a project that is of high priority to one segment of an organization is also of high priority to all groups within the organization. Guidelines for securing cooperation for projects are

- Establish achievable goals for the project
- Clearly outline individual commitments required
- Integrate project priorities with existing priorities
- Eliminate the fear of job loss due to automation
- Anticipate and eliminate potential sources of conflict
- Use an open-door policy to address project grievances
- Remove skepticism by documenting the merits of the project.

For resource management, there are different types of cooperation that should be considered and encouraged. Some are listed here:

1. *Functional Cooperation*: This is cooperation induced by the nature of the functional relationship between two groups. The two groups may be required to perform related functions that can only be accomplished through mutual cooperation.
2. *Social Cooperation*: This is the type of cooperation effected by the social relationship between two groups. The prevailing social relationship motivates cooperation that may be useful in getting project work done.
3. *Legal Cooperation*: Legal cooperation is the type of cooperation that is imposed through some authoritative requirement. In this case, the participants may have no choice other than to cooperate.

4. *Administrative Cooperation*: This is cooperation brought on by administrative requirements that make it imperative that two groups work together on a common goal.
5. *Associative Cooperation*: This is a type of cooperation that may also be referred to as collegiality. The level of cooperation is determined by the association that exists between two groups.
6. *Proximity Cooperation*: Cooperation due to the fact that two groups are geographically close is referred to as proximity cooperation. Being close makes it imperative that the two groups work together.
7. *Dependency Cooperation*: This is cooperation caused by the fact that one group depends on another group for some important aspect. Such dependency is usually of a mutually two-way nature. One group depends on the other for one thing, while the latter group depends on the former for some other thing.
8. *Imposed Cooperation*: In this type of cooperation, external agents must be employed to induce cooperation between two groups. This is applicable for cases where the two groups have no natural reason to cooperate. This is where the approaches presented earlier for seeking cooperation can become very useful.
9. *Lateral Cooperation*: Lateral cooperation involves cooperation with peers and immediate associates. Lateral cooperation is often easy to achieve because existing lateral relationships create a conducive environment for project cooperation.
10. *Vertical Cooperation*: Vertical or hierarchical cooperation refers to cooperation that is implied by the hierarchical structure of the project. For example, subordinates are expected to cooperate with their superiors.

Whichever type of cooperation is available in a project environment, the cooperative forces should be channeled toward achieving project goals. A documentation of the prevailing level of cooperation is useful for winning further support for a project. Clarification of project priorities will facilitate personnel cooperation. Relative priorities of multiple projects should be specified so that a project that is of high priority to one segment of an organization is also of high priority to all groups within the organization. Interestingly, competition can be used as a mechanism for cooperation. This happens in an environment where constructive competition paves the way for cooperation of the form that we can refer to as "*coopetition*."

PROJECT COMMITMENT

Cooperation must be supported with commitment. To cooperate is to support the ideas of a project. To commit is to willingly and actively participate in project efforts again and again throughout both the easy times and the vicissitudes of the project. Ungrudging provision of resources is one way that management can express commitment to a project.

Triple C + Commitment = Project success

By using a Pareto-type distribution, the cooperating elements of a project can be classified into three levels:

1. Top 10% (easily cooperative)
2. Middle 80% (good prospects for cooperation)
3. Bottom 10% (lost causes).

In terms of where to place efforts to get cooperation, the top 10% do not need much effort while the bottom 10% do not deserve much effort. The top 10% are the motivated individuals who will easily cooperate (i.e., Type Y workers) ,while the bottom 10% are the disagreeable individuals who will fail to see reason no matter what is presented to them (i.e., Type X workers). The best way to deal with those bottom 10% is through accommodation or exclusion. Project efforts should be concentrated where the most gains can be achieved.

PROJECT COORDINATION

After successfully initiating the communication and cooperation functions, the efforts of the project personnel must be coordinated. Coordination facilitates harmonious organization of project efforts. The development of a responsibility chart can be very helpful at this stage. A responsibility chart is a matrix consisting of columns of individual or functional departments and rows of required actions. An example of the chart is presented in Table 2.1. Cells within the matrix are filled with relationship codes that indicate who is responsible for what. The responsibility matrix helps to avoid neglecting crucial communication requirements and obligations. It can help resolve questions such as the following:

- Who is to do what?
- How long will it take?
- Who is to inform whom of what?
- Whose approval is needed for what?
- Who is responsible for which results?
- What personnel interfaces are required?
- What support is needed from whom and when?

RESOLVING PROJECT CONFLICTS

When implemented as an integrated process, the Triple C model can help avoid conflicts in a project. When conflicts do develop, it can help in resolving the conflicts. Several types of conflicts can develop in the project environment. Some of these conflicts are listed and discussed:

Scheduling Conflicts: These can develop because of improper timing or sequencing of project tasks. This is particularly common in large multiple projects. Procrastination can lead to having too much to do at once, thereby creating

TABLE 2.1
Project Responsibility Matrix

Tasks	Person responsible[a]							Status of task[b]				
								Jan	Feb	Mar	Apr	May
	A	B	C	D	E	F	G	1	1	1	1	1
1. Brainstorming meeting	R	R	R	R				D				
2. Identify speakers			R						O			
3. Select seminar location	I	R	R						O			
4. Select banquet location	R	R							D			
5. Prepare publicity materials		C	R	I	I	R						
6. Draft brochures		C	R			R						
7. Develop schedule			R									
8. Arrange for visual aids			R									
9. Coordinate activities			R									
10. Periodic review of tasks	R	R	R	S	I							
11. Monitor progress of program	C	R	R									
12. Review program process	R											
13. Closing arrangements	R											
14. Postprogram review and evaluation	R	R	R	R								

[a] Responsibility code: R (responsible), I (inform), S (support), and C (consult).
[b] Status code: D (done), O (on track), and D (delayed).

a clash of project functions and discord between project team members. Inaccurate estimates of time requirements may lead to unfeasible activity schedules. Project coordination can help avoid schedule conflicts.

Cost Conflicts: Project cost may not be generally acceptable to the clients of a project. This will lead to project conflict. Even if the initial cost of the project is acceptable, a lack of cost control during project implementation can lead to conflicts. Poor budget allocation approaches and the lack of financial feasibility study will cause cost conflicts later on in a project. Communication and coordination can help prevent most of the adverse effects of cost conflicts.

Performance Conflicts: If clear performance requirements are not established, performance conflicts will develop. A lack of clearly defined performance standards can lead each person to evaluate his or her own performance based on personal value judgments. In order to uniformly evaluate quality of work and monitor project progress, performance standards should be established by using the Triple C approach.

Management Conflicts: There must be a two-way alliance between management and the project team. The views of management should be understood by the team. The views of the team should be appreciated by management. If this does not happen, management conflicts will develop. A lack of a two-way interaction can lead to strikes and industrial actions which can be detrimental to project objectives. The Triple C approach can help create a conducive dialogue environment between management and the project team.

Technical Conflicts: If the technical basis of a project is not sound, technical conflicts will develop. Manufacturing and automation projects are particularly prone to technical conflicts because of their significant dependence on technology. Lack of a comprehensive technical feasibility study will lead to technical conflicts. Performance requirements and systems specifications can be integrated through the Triple C approach to avoid technical conflicts.

Priority Conflicts: These can develop if project objectives are not defined properly and applied uniformly across a project. Lack of direction in project definition can lead each project member to define individual goals that may be in conflict with the intended goal of a project. Lack of consistency of project mission is another potential source of priority conflicts. Overassignment of responsibilities with no guidelines for relative significance levels can also lead to priority conflicts. Communication can help defuse priority conflicts.

Resource Conflicts: Resource allocation problems are a major source of conflicts in project management. Competition for resources, including personnel, tools, hardware, software, and so on, can lead to disruptive clashes among project members. The Triple C approach can help secure resource cooperation.

Power Conflicts: Project politics can lead to power plays as one individual seeks to widen his or her scope of power. This can, obviously, adversely affect the progress of a project. Project authority and project power should be clearly differentiated: Project authority is the control that a person has by virtue of his or her functional post, while project power relates to the clout and influence that a person can exercise owing to connections within the administrative structure. People with popular personalities can often wield a lot of project power in spite of low or nonexistent project authority. The Triple C model can facilitate a positive marriage of project authority and power to the benefit of project goals. This will help define clear leadership for a project.

Personality Conflicts: These are a common problem in projects involving a large group of people. The larger a project, the larger the size of the management team needed to keep things running. Unfortunately, the larger management team also creates an opportunity for personality conflicts. Communication and cooperation can help defuse personality conflicts. Some guidelines for resolving project conflicts are as follows:

• Approach the source of conflict
• Gather all the relevant facts
• Notify those involved in writing

- Solicit mediation
- Report to the appropriate authorities
- Use grievance resolution program within the organization.

REFERENCES

Badiru, Adedeji B., "Communication, Cooperation, Coordination: The Triple C of Project Management," *Proceedings of 1987 IIE Spring Conference*, Institute of Industrial Engineers, Norcross, GA, 1987, pp. 401–404.

Hammersmith, Alan G., "Implementing a PMO—The Diplomatic Pit Bull," workshop presentation at East Tennessee PMI Chapter Meeting, January 10, 2006.

3 Time and Schedule Management

On The Flight of Time

Time flies; but it has no wings.
Time goes fast; but it has no speed.
Where has time gone? But it has no destination.
Time goes here and there; but it has no direction.
Time has no embodiment; neither flies, walks, nor goes anywhere.
Yet, the passage of time is constant.
©Adedeji Badiru, 2006

As mentioned in Chapter 2, time is of the essence of managing projects. Project management can be viewed as a tripod (or a three-legged stool) with three main components as its axes:

- Time
- Budget
- Performance.

When one leg is shorter than the others or nonexistent, the tripod cannot be used for its designed purpose. This chapter focuses on the time leg of the axis and provides examples on the importance of time management.

Time is a limited non-recyclable commodity, as evidenced by the opening poem in this chapter. Industry leaders send employees to time management training sessions and continuously preach the importance of completing reports and other activities on time. However, the one area where time management is most crucial is often overlooked and undervalued, project management. Project management, by definition, is in itself time management via milestoning of crucial tasks and critical path analysis. There are only 24 hr in a day and one of the goals of any project manager is how to most efficiently use those 24 hr. Today, the trend in industry has been to sacrifice the time portion of project management and still expect the same level of performance. This thought process is flawed and ultimately leads to the failure or under delivery of a project. An activity that normally takes 12 weeks for completion is condensed into 4 weeks. This is a time compression of more than 60%. If we were to take the tripod and reduce one of the legs by 60%, the tripod would topple over. This is the same result, in terms of performance, when activity compression occurs. As stated in Chapter 1 and 2, after an idea is generated, feasibility studies most be conducted. Analysis of time constraints should be a part of these feasibility studies. Planning, organizing, scheduling, and control all have a timing component.

Planning
   How long will engineering and design take?
   How long will R&D analysis take?
   How long will it take to build up inventory?
   What is the installation time?
   How long will equipment order take?
   Will work occur during any special times (holidays or plant shutdowns)?
Organizing
   When will the necessary administrative resources be available?
   How much time can resources commit to the project?
   When are deliverables expected?
Scheduling
   When can project be executed? Is there enough time between design and
   execution?
   When will manufacturing equipment be available?
   When will production resources be available?
   Is there a special promotion that most be worked around? How does this
   effect execution timing?
Control
   Can execution time be controlled?
   In terms of budgeting what are the time constraints?
   What is the time frame for feedback? Productivity analysis? Quality analysis?
   What is the time frame to make a Go or No-Go decision?

CPM SCHEDULING

Project scheduling is often the most visible step in the sequence of steps of project management. The two most common techniques of basic project scheduling are the critical path method (CPM) and program evaluation and review technique (PERT). The network of activities contained in a project provides the basis for scheduling the project and can be represented graphically to show both the contents and objectives of the project. Extensions to CPM and PERT include precedence diagramming method (PDM) and critical resource diagramming (CRD). These extensions were developed to take care of specialized needs in a particular project scenario. PDM technique permits the relaxation of the precedence structures in a project so that the project duration can be compressed. CRD handles the project scheduling process by using activity–resource assignment as the primary scheduling focus. This approach facilitates resource-based scheduling rather than activity-based scheduling so that resources can be more effectively utilized.

CPM network analysis procedures originated from the traditional Gantt chart, or bar chart, developed by Henry L. Gantt during World War I. There have been several mathematical techniques for scheduling activities, especially where resource constraints are a major factor. Unfortunately, the mathematical formulations are not generally practical because of the complexity involved in implementing them for realistically large projects. Even computer implementations of the complex mathematical techniques often become too cumbersome for real-time managerial applications. A basic CPM project network analysis is typically implemented in three phases:

- Network-planning phase
- Network-scheduling phase
- Network-control phase.

Network Planning: In network-planning phase, the required activities and their precedence relationships are determined. Precedence requirements may be determined on the basis of the following:

- Technological constraints
- Procedural requirements
- Imposed limitations.

The project activities are represented in the form of a network diagram. The two popular models for network drawing are the activity-on-arrow (AOA) and the activity-on-node (AON). In the AOA approach, arrows are used to represent activities, while nodes represent starting and ending points of activities. In the AON approach, conversely, nodes represent activities, while arrows represent precedence relationships. Time, cost, and resource requirement estimates are developed for each activity during the network-planning phase and are usually based on historical records, time standards, forecasting, regression functions, or other quantitative models.

Network scheduling is performed by using forward-pass and backward-pass computations. These computations give the earliest and latest starting and finishing times for each activity. The amount of "slack" or "float" associated with each activity is determined. The activity path that includes the least slack in the network is used to determine the critical activities. This path also determines the duration of the project. Resource allocation and time–cost trade-offs are other functions performed during network scheduling.

Network control involves tracking the progress of a project on the basis of the network schedule and taking corrective actions when needed. An evaluation of actual performance versus expected performance determines deficiencies in the project progress. The advantages of project network analysis are as follows:

Advantages for communication
- Clarifies project objectives
- Establishes the specifications for project performance
- Provides a starting point for more detailed task analysis
- Presents a documentation of the project plan
- Serves as a visual communication tool.

Advantages for control
- Presents a measure for evaluating project performance
- Helps determine what corrective actions are needed
- Gives a clear message of what is expected
- Encourages team interaction.

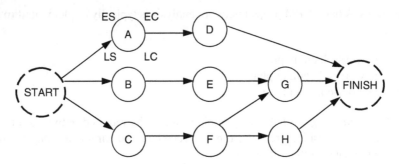

FIGURE 3.1   Graphical representation of AON network.

Advantages for team interaction
- Offers a mechanism for a quick introduction to the project
- Specifies functional interfaces on the project
- Facilitates ease of task coordination.

Figure 3.1 shows the graphical representation for AON network. The network components are

- *Node*: A node is a circular representation of an activity.
- *Arrow*: An arrow is a line connecting two nodes and having an arrowhead at one end. The arrow implies that the activity at the tail of the arrow precedes the one at the head of the arrow.
- *Activity*: An activity is a time-consuming effort required to perform a part of the overall project. An activity is represented by a node in the AON system or by an arrow in the AOA system. The job the activity represents may be indicated by a short phrase or symbol inside the node or along the arrow.
- *Restriction*: A restriction is a precedence relationship that establishes the sequence of activities. When one activity must be completed before another activity can begin, the first is said to be a predecessor of the second.
- *Dummy*: A dummy is used to indicate one event of a significant nature (e.g., milestone). It is denoted by a dashed circle and treated as an activity with zero time duration. A dummy is not required in the AON method. However, it may be included for convenience, network clarification, or to represent a milestone in the progress of the project.
- *Predecessor Activity:* A predecessor activity is one that immediately precedes the one being considered.
- *Successor Activity*: A successor activity is one that immediately follows the one being considered.
- *Descendent Activity*: A descendent activity is any activity restricted by the one under consideration.
- *Antecedent Activity*: An antecedent activity is any activity that must precede the one being considered. Activities A and B are antecedents of D. Activity A is the antecedent of B, and A has no antecedent.

- *Merge Point*: A merge point exists when two or more activities are predecessors to a single activity. All activities preceding the merge point must be completed before the merge activity can commence.
- *Burst Point*: A burst point exists when two or more activities have a common predecessor. None of the activities emanating from the same predecessor activity can be started until the burst-point activity is completed.
- *Precedence Diagram*: A precedence diagram is a graphical representation of the activities making up a project and the precedence requirements needed to complete the project. Time is conventionally shown to be from left to right, but no attempt is made to make the size of the nodes or arrows proportional to the duration of time.

ACTIVITY PRECEDENCE RELATIONSHIPS

As mentioned earlier, the precedence relationships in a CPM network fall into three major categories of technical precedence, procedural precedence, and imposed precedence. Technical precedence requirements reflect the technical relationships among activities. For example, in conventional construction, walls must be erected before the roof can be installed. Procedural precedence requirements, however, are determined by policies and procedures that may be arbitrary or subjective and may have no concrete justification. Imposed precedence requirements can be classified as resource-imposed, project-status-imposed, or environment-imposed. For example, resource shortages may require that one task be completed before another can begin, or the current status of a project (e.g., percentage of completion) may determine that one activity be performed before another, or the physical environment of a project, such as weather changes or the effects of concurrent projects, may determine the precedence relationships of the activities in a project.

The primary goal of CPM analysis is to identify the "critical path," which is a determination of the minimum completion time of a project. The computational analysis involves both forward-pass and backward-pass procedures. The forward pass determines the earliest start time and the earliest completion time for each activity in the network. The backward pass determines the latest start time and the latest completion time for each activity.

Network notations are

- A: Activity identification
- ES: Earliest starting time
- EC: Earliest completion time
- LS: Latest starting time
- LC: Latest completion time
- t: Activity duration

During the forward pass, it is assumed that each activity will begin at its earliest starting time. An activity can begin as soon as the last of its predecessors is finished. The completion of the forward pass determines the earliest completion time of the project. The backward-pass analysis is the reverse of the forward-pass analysis.

The project begins at its latest completion time and ends at the latest starting time of the first activity in the project network. Steps of CPM network analysis are presented as follows:

Step 1: Unless otherwise stated, the starting time of a project is set equal to time 0. That is, the first node, *node 1*, in the network diagram has an earliest start time of 0. Thus,

$$ES(1) = 0.$$

If a desired starting time, t_0, is specified, then $ES(1) = t_0$.

Step 2: The earliest start time (ES) for any node (activity j) is equal to the maximum of the earliest completion times (EC) of the immediate predecessors of the node. That is,

$$ES(i) = \text{Max } \{EC(j)\}$$
$$j \in P(i)$$

where $P(i) = \{$set of immediate predecessors of activity $i\}$.

Step 3: The earliest completion time (EC) of activity i is the activity's earliest start time plus its estimated time, t_i. That is,

$$EC(i) = ES(i) + t_i.$$

Step 4: The earliest completion time of a project is equal to the earliest completion time of the last node, n, in the project network. That is,

$$EC(\text{Project}) = EC(n).$$

Step 5: Unless the latest completion time (LC) of a project is explicitly specified, it is set equal to the earliest completion time of the project. This is called the zero project slack convention. That is,

$$LC(\text{Project}) = EC(\text{Project}).$$

Step 6: If a desired deadline, T_p, is specified for the project, then

$$LC(\text{Project}) = T_p.$$

It should be noted that a latest completion time or deadline may sometimes be specified for a project on the basis of contractual agreements.

Step 7: The latest completion time (LC) for activity j is the smallest of the latest start times of the activity's immediate successors. That is,

$$LC(j) = \text{Min}$$
$$i \in S(j)$$

TABLE 3.1
Data for Simple Project for CPM Analysis

Activity	Predecessor	Duration (days)
A	—	2
B	—	6
C	—	4
D	A	3
E	C	5
F	A	4
G	B, D, E	2

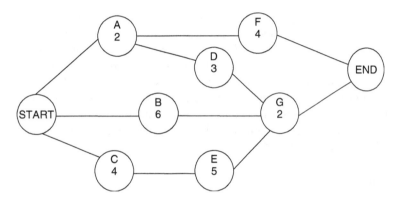

FIGURE 3.2   Example of activity network.

where $S(j)$ = {immediate successors of activity j}.
Step 8: The latest start time for activity j is the latest completion time minus the
activity time. That is,

$$LS(j) = LC(j) - t_j.$$

CPM EXAMPLE

Table 3.1 presents the data for a simple project network. The AON network for
the example is given in Figure 3.2. Dummy activities are included in the network to
designate single starting and ending points for the network.

Forward Pass

The forward-pass calculations are shown in Figure 3.3. Zero is entered as the ES for
the initial node. As the initial node for the example is a dummy node, its duration is 0.
Thus, EC for the starting node is equal to its ES. The ES values for the immediate

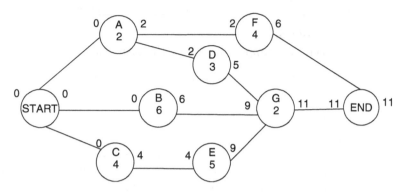

FIGURE 3.3 Forward-pass analysis for CPM example.

successors of the starting node are set equal to the EC of the START node and the resulting EC values are computed. Each node is treated as the "start" node for its successor or successors. However, if an activity has more than one predecessor, the maximum of the ECS of the preceding activities is used as the activity's starting time. This happens in the case of activity G, whose ES is determined as Max {6, 5, 9} = 9. The earliest project completion time for the example is 11 days. Note that this is the maximum of the immediately preceding earliest completion times: Max {6, 11} = 11. As the dummy ending node has no duration, its earliest completion time is set equal to its earliest start time of 11 days.

Backward Pass

The backward-pass computations establish the latest start time (LS) and latest completion time (LC) for each node in the network. The results of the backward-pass computations are shown in Figure 3.4. As no deadline is specified, the latest completion time of the project is set equal to the earliest completion time. By backtracking and using the network analysis rules presented earlier, the latest completion and latest start times are determined for each node. Note that in the case of activity A with two immediate successors, the latest completion time is determined as the minimum of the immediately succeeding latest start times. That is, Min {6, 7} = 6. A similar situation occurs for the dummy starting node. In that case, the latest completion time of the dummy start node is Min {0, 3, 4} = 0. As this dummy node has no duration, the latest starting time of the project is set equal to the node's latest completion time. Thus, the project starts at time 0 and is expected to be completed by time 11.

Within a project network, there are usually several possible paths and a number of activities that must be performed sequentially, as well as some activities that may be performed concurrently. If an activity has ES and EC times that are not equal, then the actual start and completion times of that activity may be flexible. The amount of flexibility an activity possesses is called "slack" time or "float" time. The slack time is used to determine the critical activities in the network, as will be discussed in the following sections.

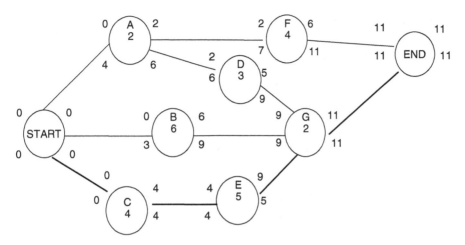

FIGURE 3.4 Backward-pass analysis for CPM example.

DETERMINATION OF CRITICAL ACTIVITIES

The critical path is defined as the path with the least slack in the network. All activities on the critical path are classified as critical activities. These activities can create bottlenecks in the project if they are delayed. The critical path is also the longest path in the network diagram. In large networks, it is possible to have multiple critical paths. In this case, it may be difficult to visually identify all the critical paths. There are four basic types of activity slack or float. They are described herewith.

- *Total Slack (TS)* is defined as the amount of time an activity may be delayed from its earliest starting time without delaying the latest completion time of the project. The total slack of activity j is the difference between the latest completion time and the earliest completion time of the activity, or the difference between the latest starting time and the earliest starting time of the activity:

$$TS(j) = LC(j) - EC(j) \text{ or } TS(j) = LS(j) - ES(j).$$

- *Free Slack (FS)* is the amount of time an activity may be delayed from its earliest starting time without delaying the starting time of any of its immediate successors. An activity's free slack is calculated as the difference between the minimum earliest starting time of the activity's successors and the earliest completion time of the activity.

$$FS(j) = Min \{EC(i)\} - EC(j)$$
$$j \in S(i)$$

- *Interfering Slack (IS)* is the amount of time by which an activity interferes with (or obstructs) its successors when its total slack is fully used. It is computed as the difference between total slack and free slack.

$$IS(j) = TS(j) - FS(j).$$

- *Independent Float (IF)* is the amount of float that an activity will always have regardless of the completion times of its predecessors or the starting times of its successors. It is computed as

$$IF = Max \{0, (Min \, ES_j - MaxLC_i - t_k)\} - EC(j)$$
$$j \in S(k), \quad i \in P(k) \quad j \in S(k) \quad i \in P(k)$$

where ES_j is the earliest starting time of the succeeding activity, LC_i is the latest completion time of the preceding activity, and t is the duration of the activity whose IF is being calculated. IF takes a pessimistic view of the situation of an activity. It evaluates the situation assuming that the activity is pressured from both sides—that is, its predecessors are delayed as late as possible while its successors are to be started as early as possible. IF is useful for conservative planning purposes. Activities can be buffered with IFs as a way to handle contingencies. For Figure 3.4, the total slack and the free slack for activity A are

$$TS = 6 - 2 = 4 \text{ days}$$
$$FS = Min \{2, 2\} - 2 = 2 - 2 = 0.$$

Similarly, the total slack and the free slack for activity F are

$$TS = 11 - 6 = 5 \text{ days}$$
$$FS = Min \{11\} - 6 = 11 - 6 = 5 \text{ days}.$$

Table 3.2 presents a tabulation of the results of the CPM example. The table contains the earliest and latest times for each activity, as well as the total and free slacks. The results indicate that the minimum total slack in the network is zero. Thus, activities C, E, and G are identified as the critical activities. The critical path is highlighted in Figure 3.4 and consists of the following sequence of activities:

$$Start \rightarrow C \rightarrow E \rightarrow G \rightarrow End.$$

The total slack for the overall project itself is equal to the total slack observed on the critical path. The minimum slack in most networks will be 0 as the ending LC is set equal to the ending EC. If a deadline is specified for a project, then the project's latest completion time should be set to the specified deadline. In that case, the minimum total slack in the network will be given by

$$TS_{Min} = (\text{Project deadline}) - EC \text{ of the last node.}$$

TABLE 3.2
Result of CPM Analysis for Sample Project

Activity	Duration (days)	ES	EC	LS	LC	TS	FS	Critical
A	2	0	2	4	6	4	0	—
B	6	0	6	3	9	3	3	—
C	4	0	4	0	4	0	0	Critical
D	3	2	5	6	9	4	4	—
E	5	4	9	4	9	0	0	Critical
F	4	2	6	7	11	5	5	—
G	2	9	11	9	11	0	0	Critical

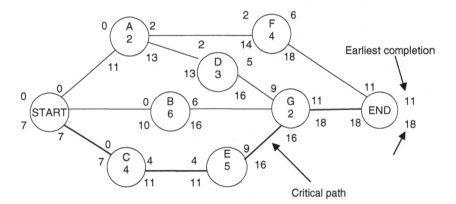

FIGURE 3.5 CPM network with deadline.

This minimum total slack will then appear as the total slack for each activity on the critical path. If a specified deadline is lower than the EC at the finish node, then the project will start out with a negative slack. That means that it will be behind schedule before it even starts. It may then become necessary to expedite some activities (i.e., crashing) in order to overcome the negative slack. Figure 3.5 shows an example with a specified project deadline. In this case, the deadline of 18 days occurs after the earliest completion time of the last node in the network.

SUBCRITICAL PATHS

In a large project network, there may be paths that are near critical. Such paths require almost as much attention as the critical path as they have a high risk of becoming critical when changes occur in the network. Analysis of subcritical paths may help in the classification of tasks into criticality categories of a A (most critical), B (medium critical), and C (least critical) categories on the basis of Pareto analysis, which separates the most important activities from the less important ones. This can be used for more targeted allocation of resources. With subcritical analysis, attention can shift from focusing only on the critical path to managing critical and near-critical tasks. Steps for identifying the subcritical paths are

Step 1: Sort activities in increasing order of total slack.
Step 2: Partition the sorted activities into groups based on the magnitudes of total slack.
Step 3: Sort the activities within each group in increasing order of their earliest starting times.
Step 4: Assign the highest level of criticality to the first group of activities (e.g., 100%). This first group represents the usual critical path.
Step 5: Calculate the relative criticality indices for the other groups in decreasing order of criticality.

Define the following variables:

α_1 the minimum total slack in the network
α_2 the maximum total slack in the network
β total slack for the path whose criticality is to be calculated

Compute the path's criticality level as:

$$\lambda = \frac{\alpha_2 - \beta}{\alpha_2 - \alpha_1} (100\%)$$

This procedure yields relative criticality levels between 0% and 100%. Table 3.3 presents an example of path criticality levels. The criticality level may be converted to a scale between 1 (least critical) and 10 (most critical) by the scaling factor:

$$\lambda' = 1 + 0.09\lambda$$

TABLE 3.3
Analysis of Subcritical Paths

Path number	Activities on path	Total slack	λ (%)	λ' (%)
1	A, C, G, H	0	100	10
2	B, D, E	1	97.56	9.78
3	F, I	5	87.81	8.90
4	J, K, L	9	78.05	8.03
5	O, P, Q, R	10	75.61	7.81
6	M, S, T	25	39.02	4.51
7	N, AA, BB, U	30	26.83	3.42
8	V, W, X	32	21.95	2.98
9	Y, CC, EE	35	17.14	2.54
10	DD, Z, FF	41	0	1.00

GANTT CHARTS

A project schedule is developed by mapping the results of CPM analysis to a calendar timeline. The Gantt chart is one of the most widely used tools for presenting project schedules. A Gantt chart can show planned and actual progress of activities. As a project progresses, markers are made on the activity bars to indicate actual work accomplished. Figure 3.6 presents the Gantt chart for our illustrative example, using the earliest starting (ES) times from Table 3.2. Figure 3.7 presents the Gantt chart for the example based on the latest starting (LS) times. Critical activities are indicated by the shaded bars.

Figure 3.6 shows that the starting time of activity F can be delayed from day two until day 7 (i.e., TS = 5) without delaying the overall project. Likewise, A, D, or both may be delayed by a combined total of 4 days (TS = 4) without delaying the overall project. If all the 4 days of slack are used up by A, then D cannot be delayed. If A is

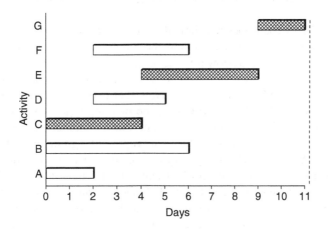

FIGURE 3.6   Gantt chart based on earliest starting times.

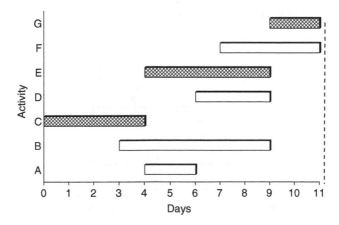

FIGURE 3.7   Gantt chart based on latest starting times.

delayed by 1 day, then D can be delayed by up to 3 days without, however, causing a delay of G, which determines project completion. The Gantt chart also indicates that activity B may be delayed by up to 3 days without affecting the project's completion time.

In Figure 3.7, the activities are shown scheduled by their latest completion times. This represents an extreme case in which activity slack times are fully used. No activity in this schedule can be delayed without delaying the project. In Figure 3.7, only one activity is scheduled over the first 3 days. This may be compared to the schedule in Figure 3.6, which has three starting activities. The schedule in Figure 3.7 may be useful if there is a situational constraint that permits only a few activities to be scheduled in the early stages of the project. Such constraints may involve shortage of project personnel, lack of initial budget, time allocated for project initiation, time allocated for personnel training, an allowance for a learning period, or general resource constraints. Scheduling of activities based on ES times indicates an optimistic view, while scheduling on the basis of LS times represents a pessimistic approach.

The basic Gantt chart does not show the precedence relationships among activities; but the chart can be modified to show these relationships by linking appropriate bars, as shown in Figure 3.8. However, the linked bars become cluttered and confusing in the case of large networks. Figure 3.9 shows a Gantt chart that presents a comparison of planned and actual schedules. Note that two tasks are in progress at the current time indicated in the figure. One of the ongoing tasks is an unplanned task. Figure 3.10 shows a Gantt chart on which important milestones have been indicated. Figure 3.11 shows a Gantt chart in which bars represent a group of related tasks that have been combined for scheduling purposes or for conveying functional relationships required on a project. Figure 3.12 presents a Gantt chart of project phases. Each phase is further divided into parts. Figure 3.13 shows a multiple-project Gantt chart. Such multiple-project charts are useful for evaluating resource allocation strategies, especially where resource loading over multiple projects may be needed for capital budgeting and cash-flow analysis decisions. Figure 3.14 shows a cumulative slippage (CumSlip) chart that is useful for project tracking and control. The chart shows the accumulation of slippage over a time span, either for individual activities or for

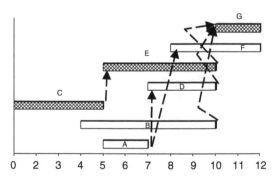

FIGURE 3.8   Linked bars in Gantt chart.

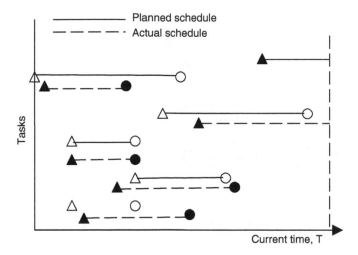

FIGURE 3.9 Progress monitoring Gantt chart.

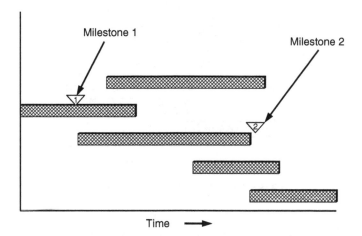

FIGURE 3.10 Milestone Gantt chart.

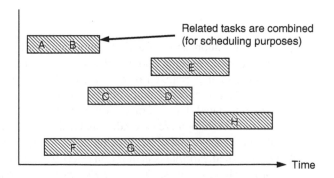

FIGURE 3.11 Task combination Gantt chart.

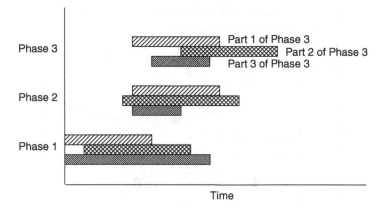

FIGURE 3.12   Phase-based Gantt chart.

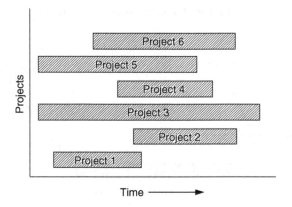

FIGURE 3.13   Multiple-projects Gantt chart.

the total project. A tracking bar is drawn to show the duration (starting and ending points) as of the last time of review.

Project Crashing

Crashing is the expediting or compression of activity duration. Crashing is done as a trade-off between a shorter task duration and a higher task cost. It must be determined whether the total cost savings realized from reducing the project duration is enough to justify the higher costs associated with reducing individual task durations. If there is a delay penalty associated with a project, it may be possible to reduce the total project cost even though crashing increases individual task costs. If the cost savings on the delay penalty is higher than the incremental cost of reducing the project duration, then crashing is justified. Normal task duration refers to the time required to perform a task under normal circumstances. *Crash task duration* refers to the reduced time required to perform a task when additional resources are allocated to it.

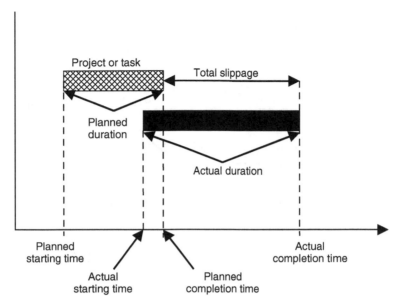

FIGURE 3.14 Cumulative slippage (CumSlip) chart.

TABLE 3.4
Normal and Crash Time and Cost Data

Activity	Normal duration (days)	Normal cost ($)	Crash duration (days)	Crash cost ($)	Crashing ratio
A	2	210	2	210	0
B	6	400	4	600	100
C	4	500	3	750	250
D	3	540	2	600	60
E	5	750	3	950	100
F	4	275	3	310	35
G	2	100	1	125	25
		$2775		$3545	

If each activity is assigned a range of time and cost estimates, then several combinations of time and cost values will be associated with the overall project. Iterative procedures are used to determine the best time or cost combination for a project. Time–cost trade-off analysis may be conducted, for example, to determine the marginal cost of reducing the duration of the project by one time unit. Table 3.4 presents an extension of the data for the example problem to include normal and crash times as well as normal and crash costs for each activity. The normal duration of the project is 11 days, as seen earlier, and the normal cost is $2775.

If all the activities are reduced to their respective crash durations, the total crash cost of the project will be $3545. In that case, the crash time is found by CPM analysis

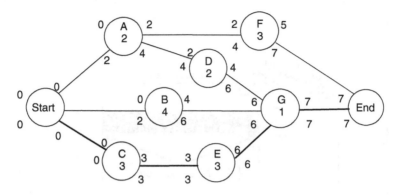

FIGURE 3.15   Example of fully crashed CPM network.

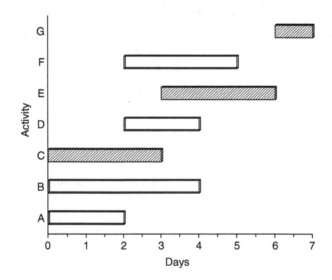

FIGURE 3.16   Gantt chart of fully crashed CPM network.

to be 7 days. The CPM network for the fully crashed project is shown in Figure 3.15. Note that activities C, E, and G remain critical. Sometimes, the crashing of activities may result in additional critical paths. The Gantt chart in Figure 3.16 shows a schedule of the crashed project using the ES times. In practice, one would not crash all activities in a network. Rather, some selection rule would be used to determine which activity should be crashed and by how much. One approach is to crash only the critical activities or those activities with the best ratios of incremental cost versus time reduction. The last column in Table 3.4 presents the respective ratios for the activities in our example. The crashing ratios are computed as

$$r = \frac{\text{Crash cost} - \text{Normal cost}}{\text{Normal duration} - \text{Crash duration}}$$

Activity G offers the lowest cost per unit time reduction of $25. If the preferred approach is to crash only one activity at a time, we may decide to crash activity G first and evaluate the increase in project cost versus the reduction in project duration. The process can then be repeated for the next best candidate for crashing, which is activity F in this case. The project completion time is not reduced any further as activity F is not a critical activity. After F has been crashed, activity D can then be crashed. This approach is repeated iteratively in order of activity preference until no further reduction in project duration can be achieved or until the total project cost exceeds a specified limit.

A more comprehensive analysis is to evaluate all possible combinations of the activities that can be crashed. However, such a complete enumeration would be prohibitive, as there would be a total of 2^c crashed networks to evaluate, where c is the number of activities that can be crashed out of the n activities in the network ($c \leq n$). For our example, only 6 out of the 7 activities in the network can be crashed. Thus, a complete enumeration will involve $2^6 = 64$ alternate networks. Table 3.5 shows 7 of the 64 crashing options. Activity G, which offers the best crashing ratio, reduces the project duration by only one day. Even though activities F, D, and B are crashed by a total of 4 days at an incremental cost of $295, they do not generate any reduction in project duration. Activity E is crashed by 2 days and generates a reduction of 2 days in project duration. Activity C, which is crashed by 1 day, generates a further reduction of 1 day in the project duration. It should be noted that the activities that generate reductions in project duration are the ones that were earlier identified as the critical activities.

In general, there may be more than one critical path; so the project analyst needs to check for the set of critical activities with the least total crashing ratio in order to minimize the total crashing cost. Also, one needs to update the critical paths every time a set of activities is crashed because new activities may become critical in the meantime. For the network given in Figure 3.15, the path C–E–G is the only critical path. Therefore, we do not need to consider crashing other jobs since the incurred cost will not affect the project completion time. There are 12 possible ways one can crash activities C, G, and E in order to reduce the project time.

TABLE 3.5
Selected Crashing Options for CPM Example

Option number	Activities crashed	Network duration (days)	Time reduction (days)	Incremental cost ($)	Total cost ($)
1.	None	11	—	—	2775
2.	G	10	1	25	2800
3.	G, F	10	0	35	2835
4.	G, F, D	10	0	60	2895
5.	G, F, D, B	10	0	200	3095
6.	G, F, D, B, E	8	2	200	3295
7.	G, F, D, B, E, C	7	1	250	3545

Several other approaches exist for determining which activities to crash in a project network. Two alternate approaches are presented below for computing the crashing ratio, r. The first one directly uses the criticality of an activity to determine its crashing ratio while the second one uses the following calculation:

$$r = \text{Criticality index}$$

$$r = \frac{\text{Crash cost} - \text{Normal cost}}{(\text{Normal duration} - \text{Crash duration})\,(\text{Criticality index})}$$

The first approach gives crashing priority to the activity with the highest probability of being on the critical path. In deterministic networks, this refers to the critical activities. In stochastic networks, an activity is expected to fall on the critical path only a percentage of the time. The second approach is a combination of the approach used for the illustrative example and the criticality index approach. It reflects the process of selecting the least-cost expected value. The denominator of the expression represents the expected number of days by which the critical path can be shortened.

4 Project Duration Diagnostics

Program evaluation review technique (PERT) incorporates variability in activity durations into project duration analysis and diagnostics. Real-life activities are often prone to uncertainties, which in turn determine the actual duration of the activities. In the critical path method (CPM), activity durations are assumed to be fixed. In PERT, the uncertainties in activity durations are accounted for by using three time estimates for each activity. The three time estimates represent the spread of the estimated activity duration. The greater the uncertainty of an activity, the wider its range of estimates will be. Predicting project completion time is essential for avoiding the proverbial "90% complete, but 90% remains" syndrome of managing large projects. The diagnostic techniques presented in this chapter are useful for having a better handle on predicting project durations.

PERT FORMULAS

PERT uses three time estimates (a, m, b), and the following equations to compute expected duration and variance for each activity:

$$t_e = \frac{a + 4m + b}{6}$$

$$s^2 = \frac{(b - a)^2}{36}$$

where
    a = optimistic time estimate
    m = most likely time estimate
    b = pessimistic time estimate $(a < m < b)$
    t_e = expected time for the activity
    s^2 = variance of the duration of the activity

After obtaining the estimate of the duration for each activity, the network analysis is carried out in the same manner as CPM analysis.

ACTIVITY TIME DISTRIBUTIONS

PERT analysis assumes that the probabilistic properties of activity duration can be modeled by the beta probability density function, which is shown in Figure 4.1 with alternate shapes. The uniform distribution between 0 and 1 is a special case of the beta distribution, with both shape parameters being equal to 1.

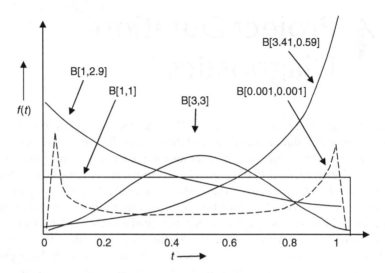

FIGURE 4.1   Beta distribution.

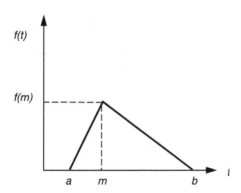

FIGURE 4.2   Triangular probability density function.

The triangular probability density function has been used as an alternative to the beta distribution for modeling activity times. The triangular density has three essential parameters: minimum value (*a*), mode (*m*), and maximum (*b*). Figure 4.2 presents a graphical representation of the triangular density function.

For cases where only two time estimates instead of three are to be used for network analysis, a uniform density function may be assumed for the activity times. This is acceptable for situations in which the extreme limits of an activity's duration can be estimated and it can be assumed that the intermediate values are equally likely to occur. Figure 4.3 presents a graphical representation of the uniform distribution. In this case, the expected activity duration is computed as the average of the upper and lower limits of the distribution. The appeal of using only two time estimates *a* and *b* is that the estimation error due to subjectivity can be reduced and the estimation task is simplified. Even when a uniform distribution is not assumed, other statistical distributions can be modeled over the range of *a* to *b*.

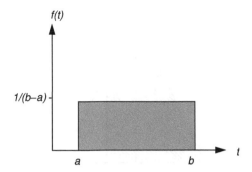

FIGURE 4.3   Uniform probability density function.

Regardless of which distribution is used, once the expected activity durations have been computed, the analysis of the activity network is carried out just as in the case of CPM.

ANALYSIS OF PROJECT DURATION

If the project duration can be assumed to be approximately normally distributed, then the probability of meeting a specified deadline can be computed by finding the area under the standard normal curve to the left of the deadline. It is important to do an analytical analysis of the project duration because project contracts often come with clauses dealing with the following:

1. No-excuse completion deadline agreement
2. Late-day penalty clause for not completing within deadline
3. On-time completion incentive award for completing on or before the deadline
4. Early-finish incentive bonus for completing ahead of schedule.

Figure 4.4 shows an example of a normal distribution describing the project duration. The variable T_d represents the specified deadline.

Using the familiar transformation formula, a relationship between the standard normal random variable z and the project duration variable can be obtained:

$$z = \frac{T_d - T_e}{S}$$

where
T_d = specified deadline
T_e = expected project duration based on network analysis
S = standard deviation of the project duration
The probability of completing a project by the deadline T_d is then computed as

$$P(T \le T_d) = P\left(z \le \frac{T_d - T_e}{S}\right)$$

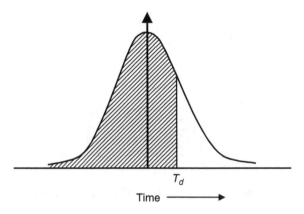

T_d

Time $\longrightarrow$

FIGURE 4.4   Project deadline under the normal curve.

TABLE 4.1
PERT Project Data

Activity	Predecessors	a	m	b	t_e	s^2
A	—	1	2	4	2.17	0.2500
B	—	5	6	7	6.00	0.1111
C	—	2	4	5	3.83	0.2500
D	A	1	3	4	2.83	0.2500
E	C	4	5	7	5.17	0.2500
F	A	3	4	5	4.00	0.1111
G	B, D, E	1	2	3	2.00	0.1111

This probability is obtained from the standard normal table available in most statistical books. Examples presented here illustrate the procedure for probability calculations in PERT. Suppose we have the project data presented in Table 4.1. The expected activity durations and variances as calculated by the PERT formulas are shown in the last two columns of the table. Figure 4.5 shows the PERT network. Activities C, E, and G are shown to be critical, and the project completion time is determined to be 11 time units.

The probability of completing the project on or before a deadline of 10 time units (i.e., $T_d = 10$) is calculated as follows:

$$T_e = 11$$
$$S^2 = V[C] + V[E] + V[G]$$
$$= 0.25 + 0.25 + 0.1111$$
$$= 0.6111$$
$$S = \sqrt{0.6111}$$
$$= 0.7817$$

$$P(T \le T_d) = P(T \le 10)$$
$$= P\left(z \le \frac{10 - T_e}{S}\right)$$
$$= P\left(z \le \frac{10 - 11}{0.7817}\right)$$
$$= P(z \le -1.2793)$$
$$= 1 - P(z \le 1.2793)$$
$$= 1 - 0.8997$$
$$= 0.1003$$

Thus, there is just over 10% probability of finishing the project within 10 days. By contrast, the probability of finishing the project in 13 days is calculated as:

$$P(T \le 13) = P\left(z \le \frac{13 - 11}{0.7817}\right)$$
$$= P(z \le 2.5585)$$
$$= 0.9948$$

This implies that there is over 99% probability of finishing the project within 13 days. Note that the probability of finishing the project in exactly 13 days will be zero. If we desire the probability that the project can be completed within a certain lower limit (T_L) and a certain upper limit (T_U), the computation will proceed as follows:

Let $T_L = 9$ and $T_U = 11.5$. Then

$$P(T_L \le T \le T_U) = P(9 \le T \le 11.5)$$
$$= P(T \le 11.5) - P(T \le 9)$$
$$= P\left(z \le \frac{11.5 - 11}{0.7817}\right) - P\left(z \le \frac{9 - 11}{0.7817}\right)$$
$$= P(z \le 0.6396) - P(z \le -2.5585)$$
$$= P(z \le 0.6396) - \left[1 - P(z \le 2.5585)\right]$$
$$= 0.7389 - [1 - 0.9948]$$
$$= 0.7389 - 0.0052$$
$$= 0.7337$$

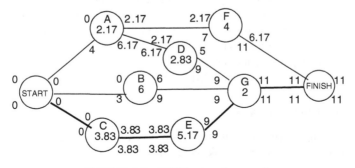

FIGURE 4.5 PERT network example.

SIMULATION OF PROJECT NETWORKS

Computer simulation is a tool that can be effectively utilized to enhance project planning, scheduling, and control. At any given time, only a small segment of a project network will be available for direct observation and analysis. The major portion of the project will either have been in the past or will be expected in the future. Such unobservable portions of the project can be analyzed through simulation analysis. Using the historical information from previous segments of a project and the prevailing events in the project environment, projections can be made about future expectations of the project. Outputs of simulation can alert management to real and potential problems. The information provided by simulation can be very helpful in project selection decisions also. Simulation-based project analysis may involve the following:

- Analytical modeling of activity times
- Simulation of project schedule
- What-if analysis and statistical modeling
- Management decisions and sensitivity analysis.

SIMULATION TOOLS

Project simulation can be built with many of the commercially available simulation packages such as SIMAN® and ARENA®. Schedule simulation is the most obvious avenue for the application of simulation to project analysis. A customized computer program called STARC is used in this section to illustrate schedule simulation and analysis. STARC is a project planning aid. It simulates project networks and performs what-if analysis of projects involving probabilistic activity times and resource constraints. The output of STARC can serve as a decision aid for project selection. The effects of different activity time estimates and resource allocation options can be studied with STARC prior to actual project commitments.

Activity Time Modeling

The true distribution of activity times will rarely be known. Even when known, the distribution may change from one type to another depending on who is performing the activity, where the activity is performed, and when the activity is performed. Simulations permit a project analyst to experiment with different activity time distributions. Commonly used activity time distributions are beta, normal, uniform, and triangular distributions. The results of simulation experiments can guide the analyst in developing definite action plans. Using the three PERT estimates a (optimistic time), m (most likely time), and b (pessimistic time), a beta distribution with appropriate shape parameters can be modeled for each activity time.

Duration Risk Coverage Factor

This is a percentage factor (q) that provides risk coverage for the potential inaccuracies in activity time estimates. A risk factor of 10% ($q = 0.10$), for example, extends

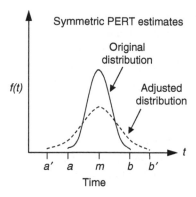

FIGURE 4.6 Interval extension for symmetric distribution.

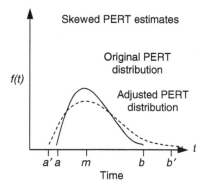

FIGURE 4.7 Interval extension for skewed distribution.

the range $[a, b]$ of an activity's duration by 10% over the specified PERT time interval. An example of the extension of the activity duration interval is shown in Figures 4.6 and 4.7. The extension of the range ensures that there is some probability (greater than zero) of generating activity times below the optimistic estimate or above the pessimistic estimate. If a risk coverage of 0% is specified, there is no interval extension and no adjustments are made to the PERT estimates. By comparison, a risk coverage of 100% yields a dilated interval that is twice as wide as the original PERT interval. Although a large extension of the PERT interval may be desirable for more simulation flexibility, it does result in a high variance for activity times.

Resource Allocation Heuristic

During a simulation run, STARC uses the composite allocation factor (CAF) to prioritize activities for resource allocation. The resource allocation process takes into account both the resource requirements and the variability in activity times. Activities with higher values of CAF are given priority during the resource allocation process. CAF is a weighted and scaled sum of two components: resource allocation factor (RAF) and stochastic activity duration factor (SAF). RAF indicates the degree of resource consumption per unit time, and SAF measures the degree of variability in activity durations. For each simulation experiment, a weighting factor, w, is used to specify the relative weights or level of importance to give to RAF and SAF in the computation of CAF. The higher the value of w on a scale of 0 to 1, the higher the importance associated with resource considerations in the scheduling process.

Simulation Examples

The sample project presented in Table 4.2 is used to illustrate project network simulation analysis using STARC. The project network is shown in Figure 4.8. The sample project consists of seven activities and one resource type. There are 10 units of the resource available.

Simulation runs were made with $w = 0.5$ and $q = 0.15$. Table 4.3 shows the output of the unconstrained PERT analysis. The expected duration (DUR), earliest start (ES), earliest completion (EC), latest start (LS), latest completion (LC), total slack (TS), free slack (FS), and critical path indicator (CRIT) are presented for each activity. The output shows that the PERT time without resource constraints is 11 months. Activities 3, 5, and 7 are on the critical path.

TABLE 4.2
Sample Project Data with Resource Constraint

Activity ID	Activity number	Preceding activities	a, m, b (in months)	Resource units required
A	1	—	1, 2, 4	3
B	2	—	5, 6, 7	5
C	3	—	2, 4, 5	4
D	4	A	1, 3, 4	2
E	5	C	4, 5, 7	4
F	6	A	3, 4, 5	2
G	7	B, D, E	1, 2, 3	6

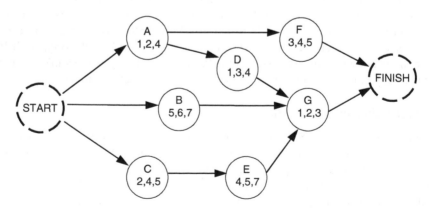

FIGURE 4.8   Project network for PERT simulation analysis.

TABLE 4.3
PERT Analysis without Resource Constraints

Activity	DUR	ES	EC	LS	LC	TS	FS	CRIT
			Unconstrained PERT schedule					
1	2.17	0.00	2.17	4.00	6.17	4.00	0.00	0.000
2	6.00	0.00	6.00	3.00	9.00	3.00	3.00	0.000
3	3.83	0.00	3.83	0.00	3.83	0.00	0.00	1.000
4	2.83	2.17	5.00	6.17	9.00	4.00	4.00	0.000
5	5.17	3.83	9.00	3.83	9.00	0.00	0.00	1.000
6	4.00	2.17	6.17	7.00	11.00	4.83	0.00	0.000
7	2.00	9.00	11.00	9.00	11.00	0.00	0.00	1.000

Unconstrained PERT project completion time = 11.00

TABLE 4.4
Simulation Output for Sample Averages

Activity	Mean DUR	Mean ES	Mean EC	Mean LS	Mean LC	Mean TS	Mean FS	CRIT Index
				Simulated sample averages				
1	2.28	6.04	8.32	6.04	8.32	0.06	0.00	0.970
2	6.04	0.00	6.04	5.17	11.21	5.17	5.11	0.000
3	3.78	0.00	3.78	2.25	6.05	2.25	0.00	0.030
4	2.81	8.32	11.13	8.40	11.21	0.14	0.08	0.840
5	5.16	3.78	8.94	6.05	11.21	2.25	2.19	0.030
6	4.01	8.32	12.33	9.22	13.23	0.91	0.00	0.130
7	2.01	11.16	13.17	11.22	13.23	0.06	0.00	0.870

Average project duration = 13.23

Table 4.4 shows the simulated sample averages for the network parameters. The average project duration is 13.23 months (based on a simulation sample size of 100). With the resource constraints, the criticality indices for activities 1 through 7 are 0.97, 0.0, 0.03, 0.84, 0.03, 0.13, and 0.87, respectively. The criticality index of an activity is the probability that the activity will fall on the critical path. The simulation indicates that activity 1 is critical most of the time (probability of 0.97), whereas activity 2 is never on the critical path.

It should be noted that the average ES, EC, LS, LC, TS, and FS will not necessarily match conventional PERT network calculations. Thus, we cannot draw a PERT network based on the sample averages. This is because each average value (e.g., mean EC) is computed from the results of several independent simulation runs. However, for each separate simulation run, the simulated outputs will fit conventional PERT network calculations.

TABLE 4.5
Project Deadline Analysis

Simulated project deadline analysis

Deadline	Calculated probability	Observed probability
10.00	0.0000	0.0000
11.00	0.0062	0.0000
12.00	0.0836	0.0700
13.00	0.3969	0.4200
14.00	0.8048	0.7900
15.00	0.9761	0.9700
16.00	0.9990	1.0000
17.00	1.0000	1.0000
18.00	1.0000	1.0000
19.00	1.0000	1.0000

Table 4.5 shows a deadline analysis for a set of selected project deadlines. The second column presents the probabilities calculated analytically on the basis of the central limit theorem. The third column presents the sample probabilities based on simulation observations. The larger the number of simulation runs, the closer the analytical and sample probabilities will be. Suppose we are considering a contract deadline of 13 months; we might like to know the probability of being able to finish the project within that time frame. The simulation output indicates a simulated probability of 0.42 for completion within 13 months. So, the chances of finishing the project in 13 months are not so good. There is a very low probability (0.0114) of finishing the project in 11 months, which is the unconstrained PERT duration. A 15-month deadline (with a probability of 0.97) seems quite achievable for this project.

Table 4.6 shows the simulation output with the shortest project duration of 11.07 months. If plotted on a Gantt chart, this schedule can serve as an operational schedule for the project, provided the simulated activity durations are realistic. It should be recalled, however, that the probability of achieving the target in 11 months is low.

The input data for the simulation analysis project network data is shown in Table 4.7. The scaled CAF measures for the activities are presented in the last column. Activity 2 has the highest priority (100.0) for resource allocation when activities compete for resources. Activity 4 has the lowest priority (54.0).

A frequency distribution histogram for the project duration based on the simulated sample is presented in Figure 4.9. As expected, the average project duration appears to be approximately normally distributed. Other portions of the simulation output show the sample duration, variances, and duration ranges. The variances and duration ranges are useful for statistical or analytical purposes such as control charts for activity durations and resource loading diagrams.

TABLE 4.6
Best Simulated Project Schedule

				Shortest schedule in simulation sample				
Activity	DUR	ES	EC	LS	LC	TS	FS	CRIT
1	1.14	5.47	6.61	5.47	6.61	0.00	0.00	1.000
2	5.47	0.00	5.47	4.31	9.78	4.31	4.16	0.000
3	4.37	0.00	4.37	0.15	4.52	0.15	0.00	0.000
4	2.04	6.61	8.65	7.74	9.78	1.13	0.98	0.000
5	5.26	4.37	9.63	4.52	9.78	0.15	0.00	0.000
6	4.46	6.61	11.07	6.61	11.07	0.00	0.00	1.000
7	1.30	9.63	10.92	9.78	11.07	0.15	0.00	0.000

Shortest simulated project duration = 11.07

TABLE 4.7
Project Simulation Data

				Project activities data			
Activity	A	M	B	Mean	Variance	Range	CAF
1	1.0	2.0	4.0	2.2	0.3	3.0	55.4
2	5.0	6.0	7.0	6.0	0.1	2.0	100.0
3	2.0	4.0	5.0	3.8	0.3	3.0	72.6
4	1.0	3.0	4.0	2.8	0.3	3.0	54.0
5	4.0	5.0	7.0	5.2	0.3	3.0	88.0
6	3.0	4.0	5.0	4.0	0.1	2.0	66.6
7	1.0	2.0	3.0	2.0	0.1	2.0	75.3

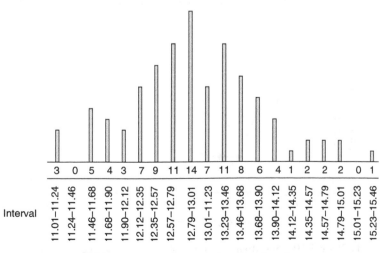

Interval	11.01–11.24	11.24–11.46	11.46–11.68	11.68–11.90	11.90–12.12	12.12–12.35	12.35–12.57	12.57–12.79	12.79–13.01	13.01–11.23	13.23–13.46	13.46–13.68	13.68–13.90	13.90–14.12	14.12–14.35	14.35–14.57	14.57–14.79	14.79–15.01	15.01–15.23	15.23–15.46
	3	0	5	4	3	7	9	11	14	7	11	8	6	4	1	2	2	2	0	1

FIGURE 4.9   Histogram of project duration.

DECISION WHAT-IF ANALYSIS

In the safe environment of simulation, the parameters involved in developing project planning strategies may be varied to determine the resulting effects on the overall project structure. Resource allocation is a major area where simulation-based what-if analysis can be useful. For example, the project analyst can determine the lower and upper bounds for resource allocation in order to achieve the specified project goal. Statistical modeling techniques, such as regression analysis, may be employed to analyze simulation data and study the relationships between factors in the project. The effectiveness of different resource allocation rules can be tested in the simulation environment.

A review of the simulation output may indicate what type of what-if analysis may be performed. For example, the number of available units of resource was increased from 10 to 15. With the additional resource units available, the average project duration was reduced from 13.23 to 11.09 months. The deadline analysis revealed that the probability of finishing the project in 13 months increased from 0.42 to 0.9895 after the additional resource allocation.

Another revision of the project data was also tested, whereby the resource availability was decreased from 10 to 7 units. It turns out that decreasing the resource availability by 3 units caused the average project duration to increase from 13.23 to 17.56 months. With the decreased resource availability, even a generous deadline of 17 months has a low probability (0.25) of being achieved. Using the revised simulation outputs, management can make better-informed decisions about resource allocation.

PROJECT SENSITIVITY ANALYSIS

Management decisions based on a simulation analysis will exhibit more validity than decisions based on absolute subjective reasoning. Simulation simplifies sensitivity analysis so that a project analyst can quickly determine what changes in project parameters will necessitate a change in management decisions. For example, with

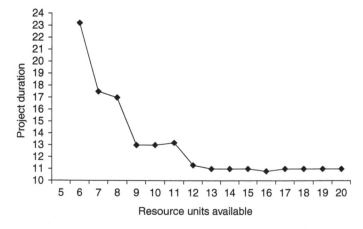

FIGURE 4.10   Sensitivity of project duration to resource availability.

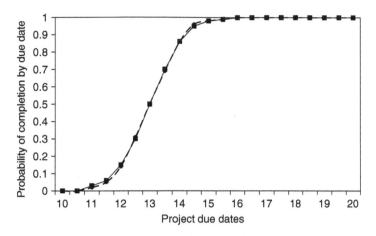

FIGURE 4.11   Sensitivity of probability of completion to project due dates.

TABLE 4.8
Project Data with Multiple Resource Constraints

Activity	Predecessor	a, m, b	Resources r_{i1}, r_{i2}
A	—	1, 2, 4	3, 0
B	—	5, 6, 7	5, 4
C	—	2, 4, 5	4, 1
D	A	1, 3, 4	2, 0
E	C	4, 5, 7	4, 3
F	A	3, 4, 5	2, 7
G	B, D, E	1, 2, 3	6, 2

Units of resource type 1 available = 10
Units of resource type 2 available = 15

the information obtained from simulation, we can study the sensitivity of project completion times to changes in resource availability. The potential effects of decisions can be studied prior to making actual resource and time commitments. For the project data presented earlier in Table 4.1, Figure 4.10 shows a plot of the sensitivity of project duration to changes in resource availability. Note that a resource level of 5 units is infeasible as there is one activity (activity G) which requires 6 units of resources. As resource units increase, the project duration decreases until it levels off at around 11 time units. Figure 4.11 shows a plot of the sensitivity of the probability of completion to changes in project due dates. The plot is based on the case of 10 resource units available. Both the analytically calculated and the simulation observed probabilities are shown. The probability of completion increases with an increase in due dates.

TABLE 4.9
Simulation Output for Project Durations

	Average project duration			
w	q = 0.0	q = 0.1	q = 0.15	q = 0.2
0.0	12.98	13.06	13.12	12.60
0.1	13.56	12.88	13.33	13.05
0.2	13.56	12.96	13.03	13.30
0.3	13.48	13.18	12.90	13.03
0.4	13.33	13.08	13.13	13.02
0.5	12.69	13.34	12.51	13.63
0.6	12.76	13.12	13.11	12.91
0.7	13.33	12.10	12.65	12.50
0.8	13.01	13.09	13.45	13.19
0.9	13.25	13.42	13.04	13.23
1.0	16.89	16.77	16.71	17.03

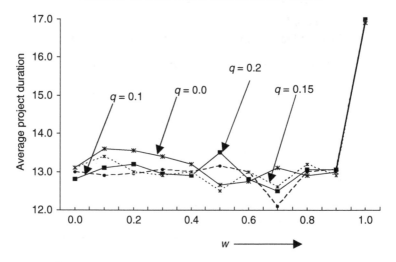

FIGURE 4.12   Plot of simulated average project duration.

MULTIPLE RESOURCE CONSTRAINTS

One additional resource type was added to the sample project data. Table 4.8 presents the revised project data. The simulation outputs for the revised project data are used for the statistical modeling presented later. Several simulation runs of the project network were made. Several combinations of w and q were investigated, and the average project durations were recorded for simulation sample size of 100. Table 4.9 presents a tabulation of the average project durations based on alternate values of w and q. Values of w range from 0.0 to 1.0, and those of q are 0, 0.1, 0.15, and 0.2.

Figure 4.12 shows a plot of the simulation output. Note that there is not much difference among the results for risk coverage levels (q) of 0%, 10%, 15%, and 20%.

Thus, the project duration appears to be insensitive to risk coverage levels less than or equal to 20%. There appear to be differences between the results for different levels of w between 0.0 and 1.0. The increase in project duration is particularly significant for values of w greater than 0.9. Thus, the project duration appears to be sensitive to changes in w.

5 Schedule Compression Techniques

Schedule compression is one way of satisfying a due date constraint on a project. We recall that any project is subject to the triple constraint of time (schedule), performance (quality), and budget (cost). One effective technique of achieving schedule compression is the precedence diagramming method (PDM), which is an extension of the basic program evaluation review technique (PERT) or critical path method (CPM). It facilitates the compression of project durations by permitting mutually dependent activities to be performed partially in parallel instead of serially. The usual finish-to-start dependencies between activities are relaxed to allow activities to be overlapped. This facilitates schedule compression. One example of a PDM relationship is to allow a successor activity to start after a certain percentage of its predecessor has been accomplished. Another PDM example is the requirement that concrete should be allowed to dry for a number of days before drilling holes for handrails. That is, drilling cannot start until so many days after the completion of concrete work. This is a finish-to-start constraint. The time between the finishing time of the first activity and the starting time of the second activity is called the lead–lag requirement between the two activities. Figure 5.1 shows the graphical representation of the basic lead–lag relationships between activity A and activity B.

SS_{AB} (Start-to-Start) Lead: This specifies that activity B cannot start until activity A has been in progress for at least SS time units.
FF_{AB} (Finish-to-Finish) Lead: This specifies that activity B cannot finish until at least FF time units after the completion of activity A.

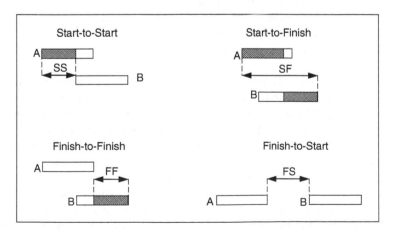

FIGURE 5.1   Lead–lag relationships in precedence diagramming.

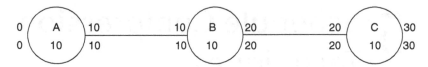

FIGURE 5.2 Set of serial activities in conventional CPM.

FS_{AB} *(Finish-to-Start) Lead*: This specifies that activity B cannot start until at least FS time units after the completion of activity A. Regular CPM and PERT analyses assume FS to be 0 for all pairs of activities.

SF_{AB} *(Start-to-Finish) Lead*: This specifies that there must be at least SF time units between the start of activity A and the completion of activity B.

The leads or lags may, alternately, be expressed in percentages rather than time units. For example, we may specify that 25% of the work content of activity A must be completed before activity B can start. If the percentage of work completed is used for determining lead–lag constraints, then a quantifiable procedure must be used for estimating the percent completion. If the project work is broken up properly using a work breakdown structure (WBS), it will be much easier to estimate percent completion by evaluating the work completed at the individual task levels. The lead–lag relationships may also be specified in terms of "*at most*" relationships instead of "*at least*" relationships. For example, we may have at most FF lag requirement between the finishing time of one activity and the finishing time of another activity. Splitting of activities often simplifies the implementation of PDM, as will be shown later with examples. Some of the factors that will determine whether or not an activity can be split include the following:

- Technical limitations affecting the splitting of a task.
- Morale of the person working on the split task.
- Setup times required to restart split tasks.
- Difficulty involved in managing resources for split tasks.
- Loss of consistency of work, and, of course.
- Existing management policy about splitting jobs.

Figure 5.2 presents a simple CPM network consisting of three activities. The activities are to be performed serially, and each has an expected duration of 10 days. The conventional CPM network analysis indicates that the duration of the network is 30 days. The earliest times and the latest times are as shown in the figure.

The Gantt chart for the example is shown in Figure 5.3. For a comparison, Figure 5.4 shows the same network but with some lead–lag constraints. For example, there is an SS constraint of 2 days and an FF constraint of 2 days between activities A and B. Thus, activity B can start as early as 2 days after activity A starts, but it cannot finish until 2 days after the completion of A. In other words, there must be *at least* 2 days between the starting times of A and B. Likewise, *at least* 2 days must separate the finishing time of A and the finishing time of B. A similar precedence relationship exists between activity B and activity C. The earliest and latest times obtained by considering the lag constraints are indicated in Figure 5.4.

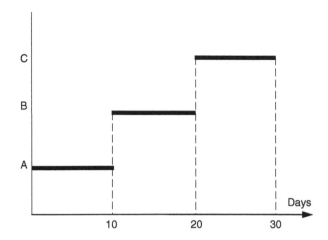

FIGURE 5.3   Gantt chart of serial activities in CPM example.

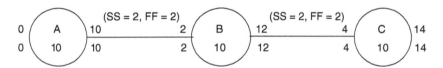

FIGURE 5.4   PDM network example.

The calculations show that if B is started just 2 days after A is started, it can be completed as early as 12 days as opposed to the 20 days obtained in the case of conventional CPM. Similarly, activity C is completed at time 14, which is considerably less than the 30 days calculated by conventional CPM. The lead–lag constraints allow us to compress or overlap activities. Depending on the nature of the tasks involved, an activity does not have to wait until its predecessor finishes before it can start. Figure 5.5 shows the Gantt chart for the example incorporating the lead–lag constraints. It should be noted that a portion of a succeeding activity can be performed simultaneously with a portion of the preceding activity.

A portion of an activity that overlaps with a portion of another activity may be viewed as a distinct portion of the required work. Thus, partial completion of an activity may be evaluated. Figure 5.6 shows how each of the three activities is partitioned into contiguous parts. Even though there is no physical break or termination of work in any activity, the distinct parts (beginning and ending) can still be identified. This means that there is no physical splitting of the work content of any activity. The distinct parts are determined on the basis of the amount of work that must be completed before or after another activity, as dictated by the lead-lag relationships. In Figure 5.6, activity A is partitioned into the parts A_1 and A_2. The duration of A_1 is 2 days because there is an SS = 2 relationship between activity A and activity B. As the original duration of A is 10 days, the duration of A_2 is then calculated to be $10 - 2 = 8$ days.

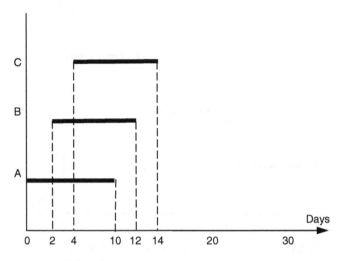

FIGURE 5.5   Gantt chart for PDM example.

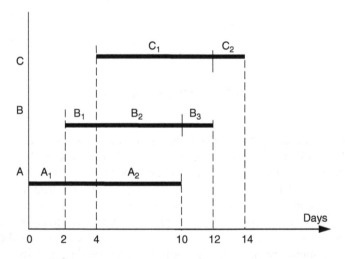

FIGURE 5.6   Partitioning of activities in PDM example.

Likewise, activity B is partitioned into parts B_1, B_2, and B_3. The duration of B_1 is 2 days because there is an SS = 2 relationship between activity B and activity C. The duration of B_3 is also 2 days because there is an FF = 2 relationship between activity A and activity B. Since the original duration of B is 10 days, the duration of B_2 is calculated to be $10 - (2 + 2) = 6$ days. In a similar fashion, activity C is partitioned into C_1 and C_2. The duration of C_2 is 2 days because there is an FF = 2 relationship between activity B and activity C. Since the original duration of C is 10 days, the duration of C_1 is then calculated to be $10 - 2 = 8$ days. Figure 5.7 shows a conventional CPM network drawn for the three activities after they are partitioned into distinct parts. The conventional forward and backward passes reveal that all the activity parts are on the critical path. This makes sense, as the original three activities are

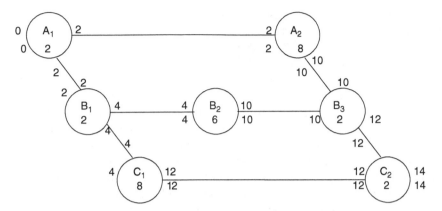

FIGURE 5.7   CPM network of partitioned activities.

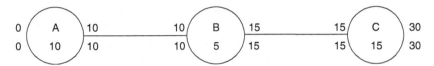

FIGURE 5.8   Another CPM example of serial activities.

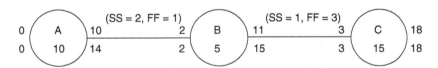

FIGURE 5.9   Compressed PDM network.

performed serially and no physical splitting of activities has been performed. Note that there are three critical paths in Figure 5.7, each with a length of 14 days. It should also be noted that the distinct parts of each activity are performed contiguously.

Figure 5.8 shows an alternate example of three serial activities. The conventional CPM analysis shows that the duration of the network is 30 days. When lead–lag constraints are introduced into the network as shown in Figure 5.9, the network duration can be compressed to 18 days.

In the forward-pass computations in Figure 5.9, note that the earliest completion time of B is time 11, because there is an FF = 1 restriction between activity A and activity B. As A finishes at time 10, B cannot finish until at least time 11. Even though the earliest starting time of B is time 2 and its duration is 5 days, its earliest completion time cannot be earlier than time 11. Also note that C can start as early as time 3 because there an SS = 1 relationship between B and C. Thus, given a duration of 15 days for C, the earliest completion time of the network is 3 + 15 = 18 days. The difference between the earliest completion time of C and the earliest completion time of B is 18 − 11 = 7 days, which satisfies the FF = 3 relationship between B and C.

In the backward pass, the latest completion time for B is 15 (i.e., $18 - 3 = 15$), as there is an $FF = 3$ relationship between activity B and activity C. The latest start time for B is time 2 (i.e., $3 - 1 = 2$), as there is an $SS = 1$ relationship between activity B and activity C. If we are not careful, however, we may erroneously set the latest start time of B to 10 (i.e., $15-5 = 10$). But that would violate the $SS = 1$ restriction between B and C. The latest completion time of A is found to be 14 (i.e., $15 - 1 = 14$), as there is an $FF = 1$ relationship between A and B. All the earliest times and latest times at each node must be evaluated to ensure that they conform to all the lead–lag constraints. When computing earliest start or earliest completion times, the smallest possible value that satisfies the lead–lag constraints should be used. By the same reasoning, when computing the latest start or latest completion times, the largest possible value that satisfies the lead-lag constraints should be used.

Manual evaluations of the lead-lag precedence network analysis can become very tedious for large networks. A software tool may be needed to simplify the implementation of PDM. If manual analysis must be done for PDM computations, it is suggested that the network be partitioned into more manageable segments. This can be done by using the WBS elements inherent in the project network. The partitioned segments may then be linked after the computations are completed. The expanded CPM network in Figure 5.10 was developed on the basis of the precedence network in Figure 5.9. It is seen that activity A is partitioned into two parts, activity B is partitioned into three parts, and activity C is partitioned into two parts. The forward and backward passes show that only the first parts of activities A and B are on the critical path. Both parts of activity C are on the critical path.

Figure 5.11 shows the corresponding earliest-start Gantt chart for the expanded network. Looking at the earliest start times, one can see that activity B is physically split at the boundary of B_2 and B_3 in such a way that B_3 is separated from B_2 by 4 days. This implies that work on activity B is temporarily stopped at time 6 after B_2 is finished and is not started again until time 10. Note that despite the 4-day delay in starting B_3, the entire project is not delayed. This is because B_3, the last part of activity B, is not on the critical path. In fact, B_3 has a total slack of 4 days. In a situation like this, the duration of activity B can actually be increased from 5 to 9 days without any adverse effect on the project duration. It should be recognized, however,

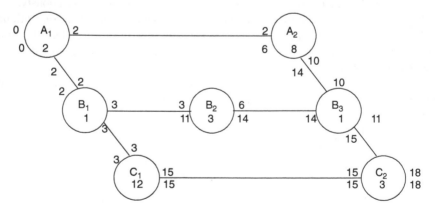

FIGURE 5.10 CPM expansion of second PDM example.

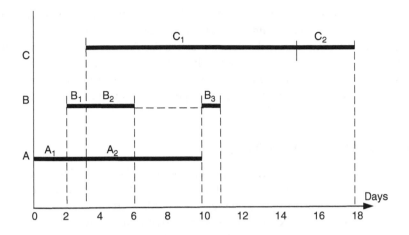

FIGURE 5.11 Compressed PDM schedule based on ES times.

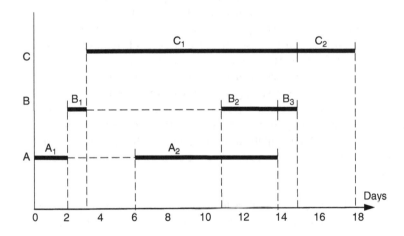

FIGURE 5.12 Compressed PDM schedule based on LS times.

that increasing the duration of an activity may have negative implications for project cost and personnel productivity.

If physical splitting of activities is not permitted, then the best option available in Figure 5.11 is to stretch the duration of B_2 so as to fill up the gap from time 6 to 10. An alternative is to delay the starting time of B_1 until time 4 so as to use up the 4-day slack right at the beginning of activity B. Unfortunately, delaying the starting time of B_1 by 4 days will delay the overall project by 4 days as B_1 is on the critical path as shown in Figure 5.10. The project analyst will need to evaluate the appropriate trade-offs between splitting of activities, delaying activities, increasing activity durations, and incurring higher project costs. The prevailing project scenario should be considered when making such trade-off decisions. Figure 5.12 shows the Gantt chart for the compressed PDM schedule based on latest start times.

In this case, it will be necessary to split both activities A and B even though the total project duration remains the same at 18 days. If activity splitting is to be avoided, then we can increase the duration of activity A from 10 to 14 days and the duration of B from 5 to 13 days without adversely affecting the entire project duration. The important benefit of precedence diagramming is that the ability to overlap activities facilitates some flexibility in manipulating individual activity times and compressing the project duration.

REVERSE CRITICAL ACTIVITIES

Working with PDM networks requires careful attention because of the potential for misinterpretation. Owing to the lead and lag requirements, activities that do not have any slacks may appear to have generous slacks. Also, "reverse-critical" activities may occur in PDM networks. Reverse-critical activities are activities that can cause a decrease in project duration when their durations are increased. This may happen when the critical path enters the completion of an activity through a finish lead-lag constraint. Also, if a "finish-to-finish" dependency and a "start-to-start" dependency are connected to a reverse-critical task, a reduction in the duration of the task may actually lead to an increase in the project duration. Figure 5.13 illustrates this unusual situation. The finish-to-finish constraint between A and B requires that B should finish no earlier than 20 days. If the duration of task B is reduced from 10 to 5 days, the start-to-start constraint between B and C forces the starting time of C to be shifted forward by 5 days, thereby resulting in a 5-day increase in the project duration.

In larger PDM networks, the preceding anomalies may occur without being noticed. One safeguard against their adverse effects is to make only one activity change at a time and document the resulting effect on the project network layout and duration. The following categorizations are often used for the unusual characteristics of activities in PDM networks:

Normal Critical (NC): This refers to an activity for which the project duration shifts in the same direction as the shift in the duration of the activity.

Reverse Critical (RC): This refers to an activity for which the project duration shifts in the reverse direction to the shift in the duration of the activity.

Bi-Critical (BC): This refers to an activity for which the project duration increases as a result of any shift in the duration of the activity.

Start Critical (SC): This refers to an activity for which the project duration shifts in the direction of the shift at the start time of the activity, but is unaffected by a shift in the overall duration of the activity.

Finish Critical (FC): This refers to an activity for which the project duration shifts in the direction of the shift at the finish time of the activity, but is unaffected by a shift in the overall duration of the activity.

Mid-Normal Critical (MNC): This refers to an activity whose middle portion is normal critical.

Mid-Reverse Critical (MRC): This refers to an activity whose middle portion is reverse critical.

Mid Bi-Critical (MBC): This refers to an activity whose middle portion is bi-critical.

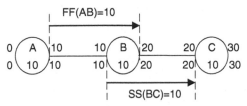

(a) Reverse critical PDM network

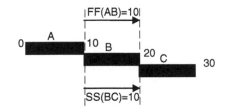

b) Gantt chart before crashing of B

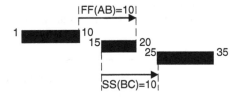

c) Gantt chart after crashing of B

FIGURE 5.13 Reverse critical activity in PDM network.

PROJECT LINE OF BALANCE

Line of balance (LOB) is a planning, scheduling, and control tool that enhances the traditional benefits of PERT/CPM and helps identify where project duration can be shortened. It is a graphical method of scheduling; and it focuses on critical activities. It identifies points that do not conform to expectations and drives corrective actions. It also encourages the *management by exception* approach to handling project problems. The main objective of LOB is to develop a report on the progress of a project. Using LOB analysis, actual progress at periodic intervals during the project can be identified. Monitoring of the flow of materials in a manufacturing system is an essential part of LOB. The requirements of LOB are to achieve the following:

- Identify objectives
- Establish a plan to meet the objectives
- Measure progress against the plan.

An objective chart is used to graphically represent project expectations. It profiles the expected level of cumulative performance toward the final project goal. This may be shown in terms of cumulative project accomplishment, total output per unit time, or cost performance. A progress chart graphically indicates the current status of the project with respect to the objectives. A production plan graphically illustrates the plan on a time basis using a logic diagram similar to a CPM chart. In production LOB, it is normal for the production plan to be the same for all identical units. The production plan provides a set of lead times for major events. For example, it may indicate the total number of time units that a task must be completed prior to the completion date of the end item. To derive the information needed to draw the line of balance, the cumulative end quantities are applied against the production plan. Thus, at any specific point in the production cycle, we can compute:

$$B = X - Y,$$

where
    B = units remaining to be completed (deficit measure for line of balance)
    X = planned units (production plan)
    Y = units completed (from progress chart).

The line of balance is drawn across the progress chart. The progress chart, thus, consists of the line of balance and the progress bars. The required quantities at a specific point in time are indicated graphically in the progress chart as a stepped horizontal line across the chart. This is the line of balance, and it identifies a given

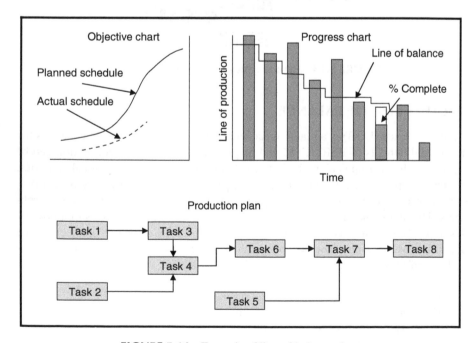

FIGURE 5.14 Example of line-of-balance chart.

position on the graph that actual quantities should reach or exceed at that measurement time. The actual quantities to date are shown graphically on the progress chart as vertical bars. It should be noted that the objective chart and the production plan typically do not change during a specific production cycle. However, the progress chart changes to reflect current level of status and must be updated at each review time. When applied to tasks, the LOB technique helps to identify tardy jobs. Figure 5.14 shows an example of a line-of-balance chart.

6 Resource Analysis and Management

This chapter addresses the issues, techniques, and tools for resource management in industrial projects. This is an important topic in project management because objectives in project management are achieved through the assignment of resources to areas of need. Consequently, resource management is essential to the achievement of successful operations. Resource management strategies, however, will vary greatly according to the nature of the specific resources that are to be managed. Some resources, such as the people who are needed to do a job, possess special skills. Other resources, whether they be of personnel, materials, or tools, may be in very limited supply. The relative importance of different resource types should be considered for resource management purposes. This chapter presents both the qualitative and quantitative aspects of resource management in manufacturing and automation projects. Topics covered in this chapter include resource allocation in project networks, resource sharing, human resource management, resource work rate analysis, *takt* time, critical-resource diagramming, and probabilistic resource utilization analysis.

RESOURCE ASSESSMENT

It is through the effective utilization of resources that project goals are achieved. Here, "resource" refers to the manpower, tools, equipment, and other physical items that are available to achieve project goals. Not all resources are necessarily tangible. Conceptual knowledge, intellectual property, and skill can all be classified as resources. The lack or untimely availability of resources is a major impediment to manufacturing and automation efforts. Thus, resource management is a complex task affected by several constraints, including the following:

- Resource interdependencies
- Conflicting resource priorities
- Mutual exclusivity of resources
- Limitations on resource availability
- Limitations on resource substitutions
- Variable levels of resource availability
- Limitations on partial resource allocation

These factors determine the tools and techniques that can be used for resource management. The assignment of tools and operators to work centers is a common resource management problem in manufacturing systems. No matter how motivated a worker is, if he or she does not have the proper tools and resources, the job cannot be done. This chapter covers both the qualitative and quantitative aspects of resource management in manufacturing and automation projects.

RESOURCE PLANNING

Project planning, in general, determines the nature of actions and responsibilities needed to achieve the project goal. It entails the development of alternate courses of action and the selection of the best action to achieve the objectives making up the goal. As resources are needed to achieve project goals, resource planning is the pivotal process that determines what needs to be done, by whom, for whom, and when. Whether it is done for long-range (strategic) purposes or short-range (operational) purposes, planning should address the following components:

1. *Project Goal and Objectives.* This planning stage involves the specification of what must be accomplished at each stage of the project. Resources constitute the primary inputs essential for achieving objectives as shown here in a flow statement moving from resources to objectives:

   Resource ➔ Activity ➔ Process ➔ Project ➔ Objectives

2. *Technical and Managerial Approach.* This stage of the planning involves the determination of the technical and managerial strategies to be employed in pursuing the project goal.
3. *Resource Availability.* This stage requires the allocation of the resources for carrying out the actions needed to achieve the project goal.
4. *Project Schedule.* This stage involves creating a logical and time-based organization of the tasks and milestones contained in the project. The schedule is typically influenced by resource limitations.
5. *Contingency Plan and Replanning.* This involves the identification of auxiliary actions to be taken in case of unexpected developments in the project.
6. *Project Policy.* This involves specifying the general guidelines for carrying out tasks within the project.
7. *Project Procedure.* This stage involves specifying the detailed method for implementing a given policy relative to the tasks needed to achieve the project goal.
8. *Performance Standard.* This stage involves the establishment of a minimum acceptable level of quality for the products of the project.
9. *Tracking, Reporting, and Auditing.* These involve keeping track of the project plans, evaluating tasks, and scrutinizing the records of the project. Figure 6.1 shows a performance-tracking chart based on resource availability.

LEVELS OF RESOURCE PLANNING

Decisions involving strategic planning lay the foundation for successful implementation of projects, as it is planning that forms the basis for all actions. Badiru (1992)

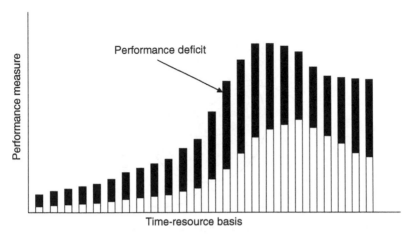

FIGURE 6.1 Resource-based performance tracking.

outlined the strategic levels of planning. Strategic decisions may be divided into three strategy levels: supra-level planning, macro-level planning, and micro-level planning.

Supra-level Planning: Planning at this level deals with the big picture of how the project fits the overall and long-range organizational goals. Questions faced at this level concern potential contributions of the project to the welfare of the organization, any possible effect on the depletion of company resources, required interfaces with other projects both within and outside of the organization, risk exposure, management support for the project, concurrent projects, company culture, market share, shareholder expectations, and financial stability.

Macro-level Planning: Planning decisions at this level address the overall planning within the boundaries of the project. The scope of the project and its operational interfaces should be addressed at this level. Questions faced at the macrolevel include goal definition, project scope, availability of qualified personnel, resource availability, project policies, communication interfaces, budget requirements, goal interactions, deadline, and conflict resolution strategies.

Micro-level Planning: This deals with detailed operational plans at the task level of the project. Definite and explicit tactics for accomplishing specific project objectives are developed at the microlevel. The concept of management by objective (MBO) may be particularly effective at this level, as MBO permits each project member to plan his or own work at the microlevel. Factors to be considered at the microlevel of project decisions include scheduled time, training requirement, tools required, task procedures, reporting requirements, and quality requirements.

The Triple C model can facilitate resource planning by identifying resource interface points and requirements for collaboration. Figure 6.2 illustrates the use of the Triple C model for resource interaction planning.

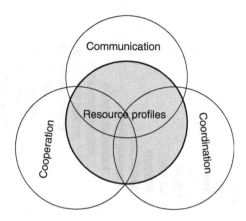

FIGURE 6.2   Resource interfaces for communication, cooperation, and coordination.

RESOURCE ALLOCATION IN PROJECT NETWORKS

Basic critical path method (CPM) and program evaluation review technique (PERT) approaches both assume theoretically unlimited resource availability in project network analysis. In this section, both the time and resource requirements of activities are considered in the development of network schedules. Projects are subject to three major constraints: time limitations, resource constraints, and performance requirements. As these constraints are difficult to satisfy simultaneously, trade-offs must be made. The smaller the resource base, the longer the project schedule. Quality of work may also be adversely affected by poor resource allocation strategies.

Good planning, scheduling, and control strategies must be developed to determine what the next desired state of a project is to be, when the next state is expected to be reached, and how to move toward that next state. Resource availability as well as other internal and external factors will determine how a project progresses from one state or stage to the next. Network diagrams, Gantt charts, progress charts, and resource loading graphs are visual aids that help determine resource allocation strategies. One of the first requirements for resource management is to determine what resources are required as opposed to what resources are available. Table 6.1 shows a model of

TABLE 6.1
Format for Resource Availability Database

Resource type	Description	Job function	When available	Duration of availability (months)	How many
Type 1	Manager	Planning	1/1/2007	10	3
Type 2	Analyst	Scheduling	2/25/2007	Indefinite	4
Type 3	Engineer	Design	Now	36	2
...	...	...	...	...	...
Type $n-1$	Operator	Machining	Immediate	Indefinite	8
Type n	Programmer	Software tools	9/2/94	12	5

a resource availability database. The database is essential when planning resource-loading strategies for resource-constrained projects.

BREAKS IN PRODUCTION SCHEDULES

In many production settings, workers encounter production breaks that require an analysis of the impact on production output. Some breaks are standard scheduled breaks while others are nonstandard and unscheduled occurrences. In each case, the workstation is subject to work rate slowdown (ramp-down) and work rate pickup (ramp-up), respectively, before and after a break. Unfortunately, production analysts typically assume that work rates remain steady up to the break time and that the scheduled work rate resumes right away after the break. This section addresses that erroneous view by introducing the the idea of work rate "ramp-down" decrease that occurs immediately before a break and the work rate "ramp-up" period that occurs immediately after the break. With this more realistic view, analysts can more accurately estimate the impacts of breaks in a production schedule.

PRODUCTION PLANNING AND TAKT TIME

Takt time refers to the production pace at which workstations must operate in order to meet a target production output rate. The production output rate is set based on product demand. As a simple example, if 2000 units of a widget are to be produced within an 8 hr shift to meet a market demand, then 250 units must be produced per hour. That means, a unit must be produced every $60/250 = 0.24$ min (14.8 sec). Thus, the takt time is 14.4 sec. Lean production planning then requires that workstations be balanced such that the production line generates a product every 14.4 sec. This is distinguished from the *cycle time*, which refers to the actual time required to accomplish each workstation task. The objective of lean production is to bring the cycle time as close to the takt time as possible. "Takt" is a German word referring to how an orchestra conductor regulates the speed, beat, or timing or the orchestra. So, the idea of takt time is to regulate (or choreograph) the rate time or pace of producing a completed product. In a balanced line design, the takt time is the reciprocal of the production rate.

Improper recognition of the role of takt time can make an analyst overestimate the production rate capability of a line. Many manufacturers have been known to overcommit to customer deliveries without accounting for the limitations imposed by takt time. As takt time is set based on customer demand, its setting may lead to an unrealistic expectations of workstations. For example, if the constraints of the prevailing learning curve will not allow sufficient learning time for new operators, then takt times cannot be sustained. This may lead to the need for buffers to temporarily accumulate units at some workstations. But this defeats the pursuits of lean production or just-in-time scheduling. The need for buffers is a symptom of imbalances in takt time. Some manufacturers build *TAKT gap* into their production planning for the purpose of absorbing nonstandard occurrences in the production line. However, if there are more nonstandard or random events than have been planned for, then production rate disruption will occur. It is important to recognize that the maximum production rate determines the minimum takt time for a production line. When demand increases, takt time should be decreased. When demand decreases, takt time should be increased. These factors as well as resource availability windows should be taken into account when scheduling industrial projects.

RESOURCE-CONSTRAINED SCHEDULING

A resource-constrained scheduling problem arises when the available resources are not enough to satisfy the requirements of activities that could be performed concurrently. To satisfy this constraint, sequencing rules (also called priority rules, activity urgency factor, scheduling rules, or scheduling heuristics) are used to determine which of the competing activities will have priority for resource allocation. Several techniques that produce an optimum are available to generating resource-constrained schedules. Unfortunately, the optimal techniques are not generally used in practice because of the complexity involved in implementing them for large projects.

Even using a computer to generate an optimal schedule is sometimes cumbersome because of the modeling requirements, the drudgery of lengthy data entry, and the combinatorial nature of interactions among activities. However, whenever possible, effort should be made in using these methods as they provide the best solution.

Most of the available mathematical techniques are based on integer programming that formulates the problem using 0 and 1 indicator variables. The variables indicate whether or not an activity is scheduled in specific time periods. Three of the common objectives in project network analysis are to minimize project duration, total project cost, and resource utilization. Whether one or more of these objectives is achieved, however, typically depends on one or more of the following constraints:

1. Limitations on resource availability
2. Precedence restrictions
3. Activity-splitting restrictions
4. Nonpreemption of activities
5. Project deadlines
6. Resource substitutions
7. Partial resource assignments
8. Mutually exclusive activities
9. Variable resource availability
10. Variable activity durations

Instead of using mathematical formulations, a scheduling heuristic uses logical rules to prioritize and assign resources to competing activities. Many such scheduling rules or heuristics have been developed in recent years.

Table 6.2 shows a project with resource constraints. There is one resource type (in this case, operators) in the project data and there are only 10 units of it available. The PERT estimates for the activity durations are expressed in terms of days. It is assumed that the resource units are reusable. Each resource unit is reallocated to a new activity at the completion of its previous assignment. Resource units can be idle

TABLE 6.2
PERT Project Data with Resource Requirements

Activity	Predecessor	a	m	b	Number of operators required
A	—	1	2	4	3
B	—	5	6	7	5
C	—	2	4	5	4
D	A	1	3	4	2
E	C	4	5	7	4
F	A	3	4	5	2
G	B, D, E	1	2	3	6

if there are no eligible activities for scheduling or if enough units are not available to start a new activity. For simplification, it is assumed that the total units of resource required by an activity must be available before the activity can be scheduled. If partial resource allocation is allowed, then the work rate of the partial resources must be determined. A methodology for determining resource work rates is presented in a subsequent section.

The unconstrained PERT duration was found earlier to be 11 days. The resource limitations are considered when creating the Gantt chart for the resource-constrained schedule. For this example, we will use the "longest-duration-first" heuristic to prioritize activities for resource allocation. Other possible heuristics are shortest-duration-first, critical-activities-first, maximum-predecessors-first, and so on. For very small project networks, many of the heuristics will yield identical schedules.

Longest-Duration-First. The initial step is to rank the activities in decreasing order of their PERT durations, This yields the following priority order:

B, E, F, C, D, A, G

At each scheduling instant, only the eligible activities are considered for resource allocation. Eligible activities are those whose preceding activities have been completed. Thus, even though activity B has the highest priority for resource allocation, it can compete for resources only if it has no pending predecessors. Referring to the PERT network shown in the preceding chapter, note that activities A, B, and C can start at time zero as none of them have predecessors. These three activities require a total of 12 operators (3 + 5 + 4) altogether. But we have only 10 operators available. So, a resource allocation decision must be made. We check our priority order and find that B and C have priority over A. So, we schedule B with 5 operators and C with 4 operators. The remaining 1 operator is not enough to meet the need of any of the remaining activities. The two scheduled activities are drawn on the Gantt chart as shown in Figure 6.3.

We have one operator idle from time 0 until time 3.83, when activity C finishes and releases 4 operators. At time 3.83, we have 5 operators available. Since activity

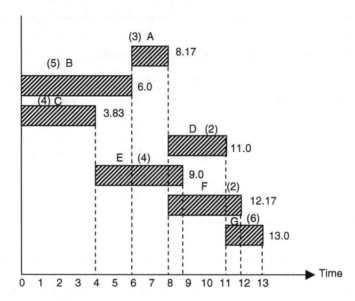

FIGURE 6.3   Resource-constrained PERT schedule.

E can start after activity C, it has to compete with activity A for resources. According to the established priority, E has priority over A. So, 4 operators are assigned to E. The remaining one operator is not enough to perform activity A, so that activity has to wait while one operator remains idle. If E had required more operators than were available, activity A would have been able to get resources at time 3.83. No additional scheduling is done until time 6, when activity B finishes and releases 5 operators. So, we now have 6 operators available and there are no activities to compete with A for resources. Thus, activity A is finally scheduled at time 6 and we are left with 3 idle operators. Even though the 3 operators are enough to start either activity D or activity F, neither of these can start until activity A finishes, because of the precedence requirement.

When activity A finishes at time 8.17, F and D are scheduled in that order. Activity G is scheduled at time 11 and finishes at time 13 to complete the project. Figure 6.3 shows the complete project schedule. The numbers in parentheses in the figure are the resource requirements. Note that we assume that activity splitting and partial resource assignments are not allowed. An activity cannot start until all the units of resources required are available. In real project situations, this assumption may be relaxed so that partial resource assignments are permissible. If splitting and partial assignments are allowed, the scheduling process will still be the same except that more recordkeeping will be required to keep track of pending jobs.

Resource allocation may be affected by several factors, including how long a resource is available (duration of availability), skill level required, cost, productivity level, and priority strategy. Ranking of activities for resource allocation may be done under either parallel priority or serial priority. In serial priority, the relative ranking of all activities is done at the beginning, before starting the scheduling process. The activities maintain their relative priority ranking throughout the scheduling process.

In the parallel priority approach, relative ranking is done at each scheduling instant and it is done only for the activities that are eligible for scheduling at that instant. Thus, under the parallel priority approach, the relative ranking of an activity may change at any time, depending on which activities it is competing with for resources at that time. This illustrative example uses the serial priority approach. If desired, any other resource allocation heuristic could be used for the scheduling example.

RESOURCE LOADING AND LEVELING

Resource profiling involves the development of graphical representations to convey information about resource availability, utilization, and assignment. Resource-loading and resource-leveling graphs are two popular tools for profiling resources. The resource-idleness graph and the critical-resource diagram are two additional tools that can effectively convey resource information.

"Resource loading" refers to the allocation of resources to work elements in a project network. A resource-loading graph presents a graphical representation of resource allocation over time. Figure 6.4 shows an example of a resource-loading graph. Separate resource-loading graphs may be drawn for the different resource types involved in a project.

The graph provides information useful for resource planning and budgeting purposes. In addition to resource units committed to activities, the graph may also be drawn for other tangible and intangible resources of an organization. For example, a variation of the graph may be used to present information about the depletion rate of the budget available for a project. If the graph is drawn for multiple resources, it can help identify potential areas of resource conflicts. For situations in which a single resource unit is assigned to multiple tasks, a variation of the resource-loading graph can be developed to show the level of load (responsibilities) assigned to the resource over time.

Resource leveling refers to the process of reducing the period-to-period fluctuation in a resource-loading graph. If resource fluctuations are beyond acceptable limits, actions can be taken to move activities or resources around in order to level out the resource-loading graph. Proper resource planning will facilitate a reasonably

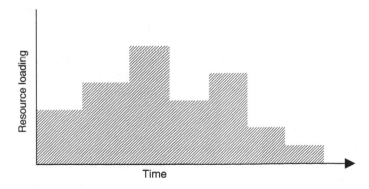

FIGURE 6.4 Resource-loading graph.

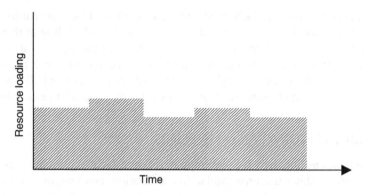

FIGURE 6.5   Resource loading after leveling.

stable level of the work force. Advantages of resource leveling include simplified resource tracking and control, lower cost of resource management, and improved opportunity for learning. Acceptable resource leveling is typically achieved at the expense of longer project duration or higher project cost. Figure 6.5 shows the leveled resource loading profile.

It should be noted that not all of the resource fluctuations in a loading graph can be eliminated. Resource leveling attempts to minimize fluctuations in resource loading by shifting activities within their available slacks. One heuristic procedure for leveling resources, known as Burgess's method, is based on the technique of minimizing the sum of the squares for the resource requirements in each period.

A resource-idleness graph is similar to a resource-loading graph except that it is drawn for the number of unallocated resource units over time. The area covered by the resource-idleness graph may be used as a measure of the effectiveness of the scheduling strategy employed for a project. Suppose two scheduling strategies yield the same project duration, and a measure of the resource utilization under each strategy is desired as a means to compare the strategies. Figure 6.6 shows two hypothetical resource-idleness graphs for the two alternate strategies. The areas are computed as follows:

$$\text{Area A} = 6(5) + 10(5) + 7(8) + 15(6) + 5(16) = 306 \text{ resource-time}$$
$$\text{Area B} = 5(6) + 10(9) + 3(5) + 6(5) + 3(3) + 12(12) = 318 \text{ resource-time}$$

As Area A is less than Area B, it is concluded that Strategy A uses resources more effectively than Strategy B. Similar measures can be developed for multiple resources to evaluate strategies for resource allocation.

CRITICAL-RESOURCE DIAGRAMMING

Badiru (1992, 1993) introduced the critical-resource diagram (CRD) as a tool for resource management. Figure 6.7 shows an example of a CRD for a project that requires six different resource types. Each node identification, RES *j*, refers to a

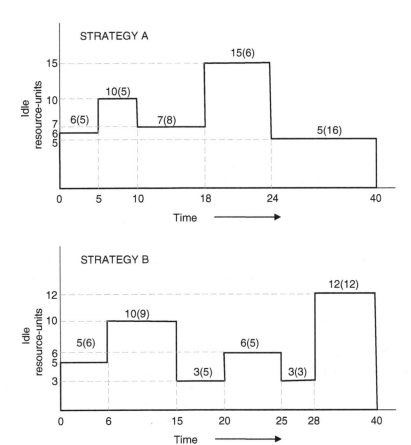

FIGURE 6.6 Resource-idleness graph.

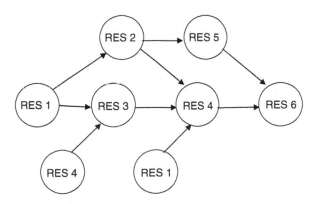

FIGURE 6.7 Basic critical-resource diagram.

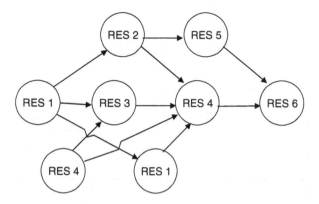

FIGURE 6.8 CRD with singular resource precedence constraint.

task responsibility for resource type *j*. In a CRD, a node is used to represent each resource unit. The interrelationships between resource units are indicated by arrows. The arrows are referred to as resource-relationship (R-R) arrows. For example, if the job of Resource 1 must precede the job of Resource 2, then an arrow is drawn from the node for Resource 1 to the node for Resource 2.

Task durations are included in a CRD to provide further details about resource relationships. Unlike in activity diagrams, a resource unit may appear at more than one location in a CRD provided there are no time or task conflicts. Such multiple locations indicate the number of different jobs for which the resource is responsible. This information may be useful for task distribution and resource-leveling purposes. In Figure 6.8, Resource Type 1 (RES 1) and Resource Type 4 (RES 4) appear at two different nodes, indicating that each is responsible for two different jobs within the same work scenario. However, appropriate precedence constraints may be attached to the nodes associated with the same resource unit if the resource cannot perform more than one task at the same time. This is illustrated in Figure 6.8.

In an application context, CRD can be used to evaluate the utilization of tools, operators, and machines in a manufacturing system. Effective allocation of these resources will improve their utilization levels. If tools that are required at several work sites are not properly managed, bottleneck problems may develop. Operators may then have to sit idle waiting for tools to become available, or an expensive tool may have to sit unused while waiting for an operator. If work cannot begin until all required tools and operators are available, then other tools and operators that are ready to work may be rendered idle while waiting for the bottlenecked resources. A CRD analysis can help identify when and where resource interdependencies occur so that appropriate reallocation actions may be taken. When there are several operators, any one operator who performs his or her job late will hold up everyone else.

CRD Network Analysis

The same forward and backward computations used in CPM are applicable to a CRD. However, the interpretation of the critical path may be different as a single

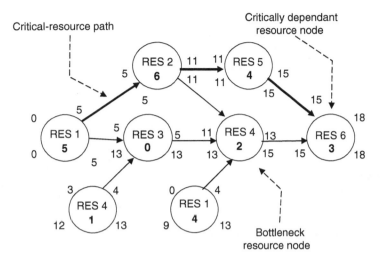

FIGURE 6.9   CRD network analysis.

resource may appear at multiple nodes. Figure 6.9 presents a computational analysis of the CRD network in Figure 6.7. Task durations (days) are given below the resource identifications. Earliest and latest times are computed and appended to each resource node in the same manner as in CPM analysis. RES 1, RES 2, RES 5, and RES 6 form the critical-resource path. These resources have no slack times with respect to the completion of the given project. Note that only one of the two tasks of RES 1 is on the critical-resource path.

Thus, RES 1 has some slack time for performing one job while it has no slack time for performing the other. None of the two tasks of RES 4 is on the critical-resource path. For RES 3, the task duration is specified as zero. Despite this favorable task duration, RES 3 may turn out to be a bottleneck resource. RES 3 may be a senior manager whose task is that of signing a work order. But if he or she is not available to sign at the appropriate time, then the tasks of several other resources may be adversely affected. A major benefit of a CRD is that both senior-level and lower-level resources can be included in the resource-planning network.

A *bottleneck* resource node is defined as a node at which two or more arrows merge. In Figure 6.9, RES 3, RES 4, and RES 6 have bottleneck resource nodes. The tasks to which bottleneck resources are assigned should be expedited in order to avoid delaying dependent resources. A *dependent* resource node is a node whose job depends on the job of immediate preceding nodes. A *critically dependent* resource node is defined as a node on the critical resource path at which several arrows merge. In Figure 6.9, RES 6 is both a critically dependent resource node as well as a bottleneck resource node. As a scheduling heuristic, it is recommended that activities that require bottleneck resources be scheduled as early as possible. A *burst* resource node is defined as a resource node from which two or more arrows emanate. As with bottleneck resource nodes, burst resource nodes should be expedited as their delay will affect several following resource nodes.

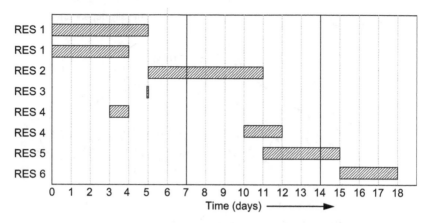

FIGURE 6.10   Resource schedule chart based on earliest start times.

Resource Schedule Chart

The CRD has the advantage that it can be used to model partial assignment of resource units across multiple tasks in single or multiple projects. A companion chart for this purpose is the resource schedule (RS) chart. Figure 6.10 shows an example of an RS chart based on the earliest times computed in Figure 6.9. A horizontal bar is drawn for each resource unit or resource type. The starting point and the length of each resource bar indicate the interval of work for the resource. Note that the two jobs of RES 1 overlap over a 4-day time period. By comparison, the two jobs of RES 4 are separated by a period of 6 days. If RES 4 is not to be idle over those 6 days, "fill-in" tasks must be assigned to it. For resource jobs that overlap, care must be taken to ensure that the resources do not need the same tools (e.g., equipment, computers, lathe, etc.) at the same time. If a resource unit is found to have several jobs overlapping over an extensive period of time, then a task reassignment may be necessary to offer some relief for the resource. The RS chart is useful for a graphical representation of the utilization of resources. Although similar information can be obtained from a conventional resource loading graph, the RS chart gives a clearer picture of where and when resource commitments overlap. It also shows areas where multiple resources are working concurrently.

RESOURCE WORK RATE ANALYSIS

When resources work concurrently at different work rates, the amount of work accomplished by each may be computed by a work rate analysis procedure. The CRD and the resource schedule chart provide information to identify when, where, and which resources work concurrently. The general relationship between work, work rate, and time can be expressed as:

$$w = rt,$$

where

w = amount of actual work accomplished (expressed in appropriate units, such as miles of road completed, lines of computer code typed, gallons of oil spill cleaned, units of widgets produced, or surface area painted.)

r = rate at which the work is accomplished

t = total time required to accomplish the work

It should be noted that work rates can change because of the effects of learning curves. In the discussions that follow, it is assumed that work rates remain constant for at least the duration of the work being analyzed. Work is defined as a physical measure of accomplishment with uniform density (i.e., homogeneous). For example, a computer programming task may be said to be homogeneous if one line of computer code is as complex and desirable as any other line of code in the program. Similarly, cleaning one gallon of oil spill is as good as cleaning any other gallon of oil spill within the same work environment. The production of one unit of a product is identical to the production of any other unit of the product. If uniform work density cannot be assumed for the particular work being analyzed, then the relationship presented here will need to be modified. If the total work to be accomplished is defined as one whole unit, then the tabulated relationship will be applicable for the case of a single resource performing the work:

Resource	Work rate	Time	Work done
Machine A	$1/x$	t	1.0

where $1/x$ is the amount of work accomplished per unit time. For a single resource to perform the whole unit of work, we must have the following:

$$(1/x)(t) = 1.0.$$

That means the magnitude of x must equal the magnitude of t. For example, if Machine A is to complete one work unit in 30 min, it must work at the rate of 1/30 of work per unit time. If the magnitude of x is greater than the magnitude of t, then only a fraction of the required work will be performed. The information about the proportion of work completed may be useful for resource planning and productivity measurement purposes. In the case of multiple resources performing the work simultaneously, the work relationship is as follows:

Resource type i	Work rate, r_i	Time, t_i	Work done, w_i
RES 1	r_1	t_1	$(r_1)(t_1)$
RES 2	r_2	t_2	$(r_2)(t_2)$
...	...	...	
RES n	r_n	t_n	$(r_n)(t_n)$

For multiple resources, we have the following expression:

$$\sum_{i-1}^{n} r_i t_i = 1.0$$

where

n = number of different resource types
r_i = work rate of resource type i
t_i = work time of resource type i

The expression indicates that even though the multiple resources may work at different rates, the sum of the total work they accomplished together must equal the required whole unit. For partial completion of work, the expression becomes:

$$\sum_{i=1}^{n} r_i t_i = p$$

where p is the proportion of the required work actually completed. Suppose RES 1, working alone, can complete a job in 50 min. After RES 1 has been working on the job for 10 min, RES 2 is assigned to help RES 1 in completing the job. Both resources working together finish the remaining work in 15 min. It is desired to determine the work rate of RES 2.

The amount of work to be done is 1.0 whole unit. The work rate of RES 1 is 1/50 of work per unit time. Therefore, the amount of work completed by RES 1 in the 10 min that the resource worked alone is $(1/50)(10) = 1/5$ of the required work. This may also be expressed in terms of percent completion or earned value using C/SCSC (Cost-Schedule Control Systems Criteria). The remaining work to be done is 4/5 of the total work. The two resources working together for 15 minutes yield the following results:

Resource type i	Work rate, r_i	Time, t_i	Work done, w_i
RES 1	1/50	15	15/50
RES 2	r_2	15	$15(r_2)$
		Total	4/5

Thus, we have $15/50 + 15(r_2) = 4/5$, which yields $r_2 = 1/30$ for the work rate of RES 2. This means that RES 2, working alone, could perform the job in 30 min. In this example, it is assumed that both resources produce identical quality of work. If quality levels are not identical for multiple resources, then the work rates may be adjusted to account for the different quality levels or a quality factor may be introduced into the analysis. The relative costs of the different resource types needed to perform the required work may be incorporated into the analysis as shown in the following table:

Resource (i)	Work rate (r_i)	Time (t_i)	Work done (w)	Ray rate (p_i)	Total cost (C_i)
Machine A	r_1	t_1	$(r_1)(t_1)$	p_1	C_1
Machine B	r_2	t_2	$(r_2)(t_2)$	p_2	C_2
...	...	...	...	...	...
Machine n	r_n	t_n	$(r_n)(t_n)$	p_n	C_n
		Total	1.0		Budget

As another example, suppose the work rate of RES 1 is such that it can perform a certain task in 30 days. It is desired to add RES 2 to the task so that the completion time of the task could be reduced. The work rate of RES 2 is such that this resource can perform the same task alone in 22 days. If RES 1 has already worked 12 days on the task before RES 2 comes in, find the completion time of the task. It is assumed that RES 1 starts the task at time 0.

As usual, the amount of work to be done is 1.0 whole unit (i.e., the full task). The work rate of RES 1 is 1/30 of the task per unit time and the work rate of RES 2 is 1/22 of the task per unit time. The amount of work completed by RES 1 in the 12 days the resource worked alone is $(1/30)(12) = 2/5$ (or 40%) of the required work. Therefore, the remaining work to be done is 3/5 (or 60%) of the full task. Let T be the time for which both resources work together. The two resources working together to complete the task yield the following table:

Resource type i	Work rate, r_i	Time, t_i	Work done, w_i
RES 1	1/30	T	$T/30$
RES 2	1/22	T	$T/22$
		Total	3/5

Thus, we have $T/30 + T/22 = 3/5$, which yields $T = 7.62$ days. Consequently, the completion time of the task is $(12 + T) = 19.62$ days from time 0. The results of this example are summarized in the resource schedule charts in Figure 6.11. It is assumed that both resources produce identical quality of work and that the respective work rates remain consistent. As mentioned earlier, the respective costs of the different resource types may be incorporated into the work rate analysis.

CASE EXAMPLE OF CRD APPLICATION

The real-case project presented in Table 6.3 is used to illustrate a project network analysis using CRD and CPM. The project network is shown in Figure 6.12. The CPM

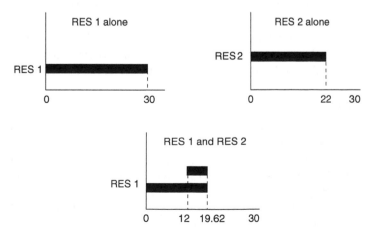

FIGURE 6.11 Resource schedule charts for RES 1 and RES 2.

TABLE 6.3
Project Data for Case Example

| Activity number | Activity name | Predecessor | Activity duration | | Resources Usage |
			Days (a, m, b)	PERT/CPM	
1	A		2, 4, 6	4	1
2	B	A	27, 30, 33	30	2→1→3
3	C	B	22, 27, 32	27	1→2
4	D	C	12, 16, 20	16	1
5	E	D	15, 17, 19	17	1→4
6	F	E	42, 44, 46	44	1→4

TABLE 6.4
Resource Availability Data for Case Example

Resource ID	Description	When available	Duration of availability
1	Project management	Immediate	Indefinite
2	Architecture analyst	Immediate	Indefinite
3	Procurement	Immediate	Indefinite
4	Deployment resources	Immediate	Indefinite

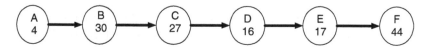

FIGURE 6.12   Project network for case example.

time estimates for each activity are shown below the activity label in the network, the time unit used is days. The sample project contains six activities and four resources type. The availability of the different resource type is presented in Table 6.4.

CPM Analysis

The forward and backward computations used in CPM were applied to this project and the result are presented in Figure 6.13.

In this project, there is no slack time, and thus the critical path contains all the activities. The duration of the project is 138 days . Figure 6.14 represents the resource loading on this project.

One can note that the RES 1 is used throughout the project. On the other hand, RES 2, RES 3, and RES 4 are used at some specific periods during the project.

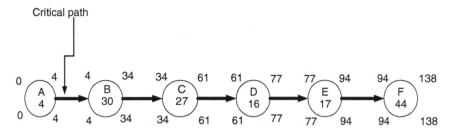

FIGURE 6.13   CPM analysis for case example.

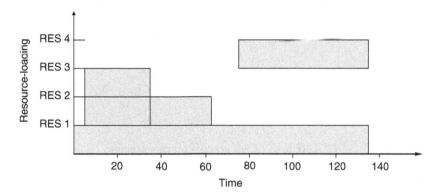

FIGURE 6.14   Resource-loading graph for case example.

CRD Analysis

Similar to the CPM, the forward and backward computations were performed on the CRD network. Figure 6.15 presents the computational analysis.

RES 1, RES 2, RES 3, and RES 4 form the critical-resource path. They have no slack times with respect to the completion of the given activity. RES 1 is a major resource in the completion of this project. Using the CRD analysis, the estimated completion time of this project is reduced to 128 days (Figure 6.16).

LINE BALANCING

Line balancing involves adjusting tasks at workstations so that it takes about the same amount of time to complete the work at each station. Most production facilities involve an integrated collection of workstations. Line balancing helps control the output rates in continuous production systems. As work on a product is completed at one station, it is passed on to the next for further processing. Cycle time is the time the product spends at each workstation. The cycle time is dependent on the expected output from the line. The cycle time can be calculated based on the production rate. A balanced line refers to the equality of output of each successive workstation on the assembly line. The maximum output of the line is controlled by its slowest operation. Perfect balance exists when each workstation requires the same amount of time

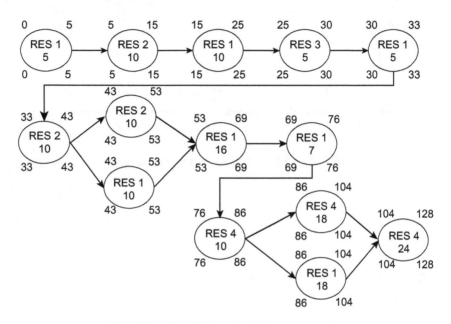

FIGURE 6.15   CRD analysis for case example.

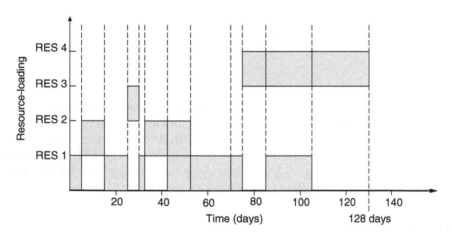

FIGURE 6.16   Resource schedule chart for case example.

to perform its assigned tasks and there is no idle time at any of the workstations. However, because of bottlenecks in operations and different work rates, perfect balance is rarely achieved.

The critical-resource diagramming approach can be effective in line-balancing analysis. The work rate table can identify specific work rates that may need to be adjusted to achieve line balance within a specified margin. The margin, in this case, is defined as the time deviation from perfect balance. The following terms are important for line-balancing analysis:

1. *Workstation*: The immediate work environment of an operator, providing all the tools necessary to perform the tasks assigned to the operator
2. *Work Element*: The smallest unit of work that cannot be divided between two or more operators without having schedule or resource conflict
3. *Operation*: A set of work elements assigned to a single workstation
4. *Cycle Time*: The time one unit of the product spends at each workstation
5. *Balance Delay*: The total idle time in the line as a result of unequal division of work between workstations

Line balancing involves grouping work elements into stations so as to minimize the number of stations based on a desired output rate or cycle time. A precedence diagram is often used to determine the logical basis for grouping work elements. The CRD can be used for this purpose. As there will be a large number of possible ways to group work elements, the analyst must use precedence constraints to eliminate the infeasible groupings.

Table Assemble (TA) conveyor system is the simplest type of line balancing. The conveyor moves work elements past workstations at a constant rate. Each station is expected to perform its task within the time period allowed by the conveyor speed. A sign of imbalance occurs when work piles up in a particular station because more units of the product arrive at the station before the preceding ones are dispatched. In this case, the next operator in the line will experience some idle time while waiting for work to arrive from the preceding station.

The choice of cycle time depends on the desired output rate. The minimum number of workstations is the total work element duration divided by the cycle time. Fractional results are rounded up to the next higher integer value. This is calculated as:

$$n = \frac{\sum_{i=1}^{k} t_i}{C}$$

where
k = number of work elements
t_i = processing time of work element i
C = cycle time

In most cases, the theoretical minimum number of workstations will be impossible to achieve because of physical constraints in the production line and/or work element times that are not compatible. It should be noted that even the theoretical minimum number of stations is not necessarily devoid of idle times. A balance with the theoretical minimum number of workstations will contain the least total idle time.

The maximum possible efficiency with the theoretical minimum number of stations is represented as:

$$f_{max} = \frac{\sum_{i=1}^{k} t_i}{nC} .$$

For cases in which the theoretical minimum number of stations cannot be achieved, the actual efficiency is computed by substituting the actual number of stations, m, for n in the above equation. As the actual efficiency, f_a, will be less than or equal to f_{max}, the analyst can attempt to increase efficiency towards the maximum value by rearranging work elements. Several mathematical and heuristic methods are used for investigating the rearrangements. These include linear programming, dynamic programming, computer simulation, trial-and-error, and the ranked positional weight technique.

HUMAN RESOURCE ISSUES

Human resources make projects successful. Human resources are distinguished from other resources because of their ability to learn, adapt to new project situations, and set goals. But a harmonious balance of human resources, technology resources, and management resources must exist in order for project goals to be successfully achieved. Managing human resources well involves placing the right people with the right skills in the right jobs in the right work environment, thus motivating workers to perform better. In order to improve the overall quality of human resources, both individual and organizational improvements are often needed. Management can create a climate for motivation by enriching jobs with the following strategies:

- Specify project goals in unambiguous terms
- Encourage and reward creativity on the job
- Eliminate mundane job-control processes
- Increase accountability and responsibility for project results
- Define jobs in terms of manageable work packages that help identify line of responsibility
- Grant formal authority to make decisions at the task level
- Create advancement opportunities in each job
- Give challenging assignments that enable a worker to demonstrate his or her skill
- Encourage upward (vertical) communication of ideas
- Provide training and tools needed to get the job done
- Maintain a stable management team

Several management approaches can be used to manage human resources. Some of these approaches are formulated as direct responses to the cultural, social, family, or religious needs of workers. Examples of these approaches are

- Flextime opportunities
- Observance of religious holidays
- Half-time employment opportunities

These approaches can have a combination of several advantages, some favoring the employer and others favoring the workers. The advantages are

- Low cost
- Cost savings on personnel benefits
- Higher employee productivity
- Less absenteeism
- Less work stress
- Better family or domestic situations, which may have positive effects on productivity

In addition to the preceding considerations, workforce retraining is important for projects involving automation. Continuing-education programs should be developed to allow workers to develop and enhance their skills. The retraining will create a ready pool of human resources that can help boost manufacturing output and competitiveness. Another consideration is that management should remain as stable as possible in order to encourage workers to adapt to the changes in industry. If management changes too often, workers may not develop a sense of commitment to the policies of management.

The major resource in any organization is manpower, both technical and nontechnical. Personnel are the overriding factor in any project life cycle. Even in automated operations, the role played by whatever few people are involved can be very significant. Such operations invariably require the services of technical people with special managerial and professional skills. The high-tech manager in such situations needs special skills in order to discharge the managerial duties effectively. The manager must have personal (automanagement) skills that relate to the following:

- Managing self
- Being managed
- Managing others

Many of the managers who supervise technical people have risen to their posts and are former technical people themselves. Consequently, they often lack the managerial competence needed at the higher levels. In some cases, technical professionals are promoted to managerial levels and then transferred away from their areas of technical expertise into administrative posts in different functional areas. If such managers perform poorly, it is not necessarily that they are incompetent; they may simply lack knowledge about the work elements in their new function. Any technical training without some management exposure is, in effect, an incomplete education. Technical professionals should be trained for the eventualities of their professions.

In the transition from the technical to the management level, an individual's attention should shift from detail to overview, from specific to general, and from technical to administrative. As most managerial positions are earned based on qualifications (except in aristocratic and autocratic systems), it is important to train technical professionals for possible administrative jobs. Therefore, it is the responsibility of both the individual and the training institution to map out career goals and paths and to institute specific education aimed at the realization of those goals. One such path is outlined in the following list; it illustrates a plausible series of steps along the way in the career trajectory for a technical professional who aspires eventually to work at the highest levels of management.

1. The *technical professional* has practical and technical training and or experience in a given field, such as industrial engineering. The technical professional must keep current in his or her area of specialization through continuing education courses, seminars, conferences, and so on. A mentoring program such as those that are now used in many large organizations can be effectively utilized at this stage of the career ladder to help a person at this level develop professionally.
2. The *project manager* has the direct responsibility of supervising a given project through the phases of planning, organizing, scheduling, monitoring, and control. Often, the managerial assignment may be limited to just a specific project and, at the conclusion of the project, the individual returns to regular technical duties. However, this person's performance on the project may help identify him or her as a suitable candidate for permanent managerial assignment later on; so, even temporary assignments may be valuable both to the organization and to the individual.
3. The *group manager* has direct responsibility for planning, organizing, and directing the activities of a group of people with a specific responsibility such as, for example, a computer data security advisory committee. Typically, this is an ongoing responsibility that may repeatedly require the managerial skills of the individual.
4. The *director* oversees a particular function of the organization. For example, a marketing director has the responsibility of developing and implementing the strategy for getting the organization's products to the right market, at the right time, at the appropriate price, and in the proper quantity. This is a critical responsibility that may directly affect the survival of the organization. Only individuals who have successfully proven themselves at the earlier career stages, ideally, get the opportunity to advance to the director's level.
5. The *administrative manager* oversees the administrative functions and staff of the organization. His or her responsibilities cut across several functional areas, requiring this person to have managerial skills and versatility that have been, again ideally, proven in previous assignments.

The positions outline here, when seen as a succession of posts, are only one of several possible paths that can be charted for a technical professional as he gradually makes the transition from the technical ranks to the management level. To function effectively, a manager must have mastered technical skills, as well as acquired nontechnical background in various subjects. Moreover, quantifiable attributes such as training and experience, and nonquantifiable ones such as attitude and personality, will all contribute to and determine a person's managerial style. A good manager's appreciation of the human and professional needs of his subordinates will substantially enhance his managerial performance. Examples of subject areas in which a manager or an aspiring manager should get training include the ones outlined in the following list. They are all skills that the effective manager will need to have mastered.

1. Project Management
   (a) *Scheduling and Budgeting*: Knowledge of project planning, organizing, scheduling, monitoring, and controlling under resource and budget restrictions
   (b) *Supervision*: Skill in planning, directing, and controlling the activities of subordinates
   (c) *Communication*: Skill of relating to others both within and outside the organization. This includes written and oral communication skills

2. Personal and Personnel Management
   (a) *Professional Development*: Leadership roles played by participating in professional societies and peer recognition acquired through professional services
   (b) *Personnel Development*: Skills needed to foster cooperation and encouragement of staff with respect to success, growth, and career advancement
   (c) *Performance Evaluation*: Development of techniques for measuring, evaluating, and improving employee performance
   (d) *Time Management*: Ability to prioritize and delegate activities as appropriate in order to maximize accomplishments within given time periods

3. Operations Management
   (a) *Marketing*: Skills useful for winning new business for the organization or preserving existing market shares
   (b) *Negotiating*: Skills for moderating personnel issues, representing the organization in external negotiations, or administering company policies
   (c) *Estimating and Budgeting*: Skills needed to develop reasonable cost estimates for company activities and the assignment of adequate resources to operations
   (d) *Cash Flow Analysis*: An appreciation for the time value of money, manipulations of equity and borrowed capitals, stable balance between revenues and expenditures, and maximization of returns on investments
   (e) *Decision Analysis*: Ability to choose the direction of work by analyzing feasible alternatives

A technical manager can develop these skills through formal college courses, seminars, workshops, short courses, professional conferences, or in-company training. Many companies appreciate the need for these skills and are willing to bear the cost of furnishing their employees with the means of acquiring the skills. Many companies even have custom-designed formal courses that they contract out to colleges to teach for their employees. This is a unique opportunity for technical professionals to acquire the managerial skills needed to move up the company ladder.

Technical people have special needs. Some of these needs, unfortunately, are often not recognized by peers, superiors, or subordinates. Inexperienced managers are particularly prone to the mistake of not distinguishing between technical and nontechnical professional needs. In order to perform more effectively, a manager must be administratively adaptable, understanding such diverse areas as the natural desire of technical professionals to stay within their area of expertise (professional preservation), the relationships among employees and their professional peers, Maslow's hierarchy of needs as it relates to the satisfaction of employees' needs, appropriateness of the technical work content to each employee, and issues relating to the quality of the managerial administration.

PROFESSIONAL PRESERVATION

Professional preservation refers to the desire of a technical professional to preserve his or her identification with a particular job function. In many situations, however, accommodating this desire is not possible because of a lack of manpower to fill specific job slots. It is common to find people trained in one technical field holding assignments in other fields. Managerially, however, there are pitfalls to this common practice. An incompatible job function, for example, can easily become the basis for insubordination, egotism, and rebellious attitudes. While it is realized that in any job environment there will sometimes be the need to work outside one's profession, every effort should be made to match the employee to the surrogate profession as closely as possible. This is primarily the responsibility of the human resources manager.

After a personnel team has been selected in the best possible manner, a critical study of the job assignments should be made. Even between two dissimilar professions, there may be specific job functions that are compatible. These should be identified and used in the process of personnel assignment. In fact, the mapping of job functions needed for an operation can serve as the basis for selecting a project team. In order to preserve the professional background of technical workers, their individualism must be understood. In most technical training programs, the professional is taught how to operate in the following manners:

1. Make decisions based on the assumption of certainty of information
2. Develop abstract models to study the problem being addressed
3. Work on tasks or assignments individually
4. Quantify outcomes
5. Pay attention to exacting details
6. Think autonomously
7. Generate creative insights to problems
8. Analyze systems operability rather than profitability

However, in the business environment, not all of the preceding characteristics are desirable or even possible. For example, many business decisions are made with incomplete data. In many situations, it is unprofitable to expend the time and efforts to seek perfect data. As another example, many operating procedures are guided by company policies rather than creative choices of employees. An effective manager should be able to spot cases where a technical employee may be given room to practice his professional training. The job design should be such that the employee can address problems in a manner compatible with his professional training.

PROFESSIONAL PEERS

In addition to having professionally compatible job functions, technical people like to have other project team members to whom they can relate technically. A project team consisting of members from diversely unrelated technical fields can be a source of miscommunication, imposition, or introversion. The lack of a professional associate on the same project can cause a technical person to fall prey to one or more of the following destructive attitudes:

1. Withdrawing into a shell and contributing very little to the project by holding back ideas that he feels the other project members cannot appreciate
2. Exhibiting technical snobbery and holding the impression that only he or she has the know-how for certain problems
3. Straddling the fence on critical issues and developing no strong conviction for project decisions

Providing an avenue for a technical "buddy system" to operate in an organization can be very helpful in assuring the congeniality of personnel teams and in facilitating the eventual success of project endeavors. The manager in conjunction with the selection committee (if one is used) must carefully consider the mix of the personnel team on a given project. If it is not possible or desirable to have more than one person from the same technical area on the project, an effort should be made to provide as good a mix as possible. It is undesirable to have several people from the same department taking issues against the views of a lone project member from a rival department. Whether it is realized or not, whether it is admitted or not, there is a keen sense of rivalry among technical fields. Even within the same field, there are subtle rivalries between specific functions. It is important not to let these differences carry over to a project environment.

WORK CONTENT

As new technology develops, the elements of a project task will need to be designed to take advantage of it. Technical professionals have a sense of achievement relative to their expected job functions. They will not be satisfied with mundane project requirements. They look forward to challenging technical assignments that will bring forth their technical competence. They prefer to claim contribution mostly where technical contribution can be identified. The project manager will need to ensure that the technical people on a project have assignments for which their background is really needed. It will be counterproductive in the end to select a technical professional for a project mainly on the basis of personality. Objective selection and an appropriate assignment of tasks will alleviate potential motivational problems that could develop later in the project.

HIERARCHY OF NEEDS

Recalling Maslow's hierarchy of needs, the needs of a technical professional should be critically analyzed. Being professionals, technical people are more likely to be higher up in the needs hierarchy. Most of their basic necessities for a good material life have already been met. Their prevailing needs will tend to involve esteem and self-actualization. As a result, by serving on a project team, a technical professional may have expectations that cannot usually be quantified in monetary terms. This is in contrast to nontechnical people, who may look forward to overtime pay or other monetary gains that may result from being on the project. Technical professionals will generally look forward to one or several of the following opportunities.

1. *Professional Growth and Advancement*: Professional growth is a primary pursuit of most technical people. For example, a computer professional has to be frequently exposed to challenging situations that introduce new technology developments and enable him to keep abreast of his field. Even occasional drifts from the field may lead to the fear of not keeping up and being left behind. The project environment must be reassuring to the technical people with regard to the opportunities for professional growth in terms of developing new skills and abilities.

2. *Technical Freedom*: Technical freedom, to the extent permissible within the organization, is essential for the full utilization of a technical background. A technical professional generally expects to have the liberty of determining how best the objective of his assignment can be accomplished. One should never impose a work method on a technical professional with the injunction that "this is the way it has always been done and will continue to be done!" If the worker's creative input to the project effort is not needed, then there is no need having him or her on the team in the first place.

3. *Respect for Personal Qualities*: Technical people have profound personal feelings despite the mechanical or abstract nature of their job functions. They will expect to be respected for their personal qualities. In spite of frequently operating in professional isolation, they do engage in interpersonal activities. They want their nontechnical views and ideas to be recognized and evaluated based on merit. They do not want to be viewed as "all technical." An appreciation for their personal qualities gives them a sense of belonging and helps them to become productive members of a project team.

4. *Respect for Professional Qualifications*: A professional qualification usually takes several years to achieve and is not likely to be compromised by any technical professional. Technical professionals cherish the attention they receive due to their technical background. They expect certain preferential treatments. They like to make meaningful contributions to the decision process. They take approval of their technical approaches for granted. They believe they are on a project because they are qualified to be there. The project manager should recognize these situations and avoid the unproductive bias of viewing the technical person as being self-conceited, instead, understanding that the well-qualified technical person may simply need to be handled with some care.

5. *Increased Recognition*: Increased recognition is expected as a by-product of a project effort. The technical professional, consciously or subconsciously,

views his participation in a project as a means of satisfying one of his higher-level needs. He or she expects to be praised for the success of his or her efforts. He or she looks forward to being invited for subsequent technical endeavors. He or she savors hearing the importance of his or her contribution being related to his peers. Without going to the extreme, the project manager can ensure the realization of the above needs through careful comments.

6. *New and Rewarding Professional Relationship*: New and rewarding professional relationships can serve as a bonus for a project effort. Most technical developments result from joint efforts of people that share closely allied interests. Professional allies are most easily found through project groups. A true technical professional will expect to meet new people with whom he can exchange views, ideas, and information later on. The project atmosphere should, as a result, be designed to be conducive to professional interactions.

QUALITY OF LEADERSHIP

The professional background of the project leader should be such that he or she commands the respect of technical subordinates. The leader must be reasonably conversant with the base technologies involved in the project. The leader must be able to converse intelligently on the terminologies of the project topic and able to convey the project ideas to upper management. This serves to give him or her technical credibility. If technical credibility is lacking, the technical professionals on the project might view him or her as an ineffective leader. They will consider it impossible to serve under a manager to whom they cannot relate technically.

In addition to technical credibility, the manager must also possess administrative credibility. There are routine administrative matters that are needed to ensure a smooth progress for the project. Technical professionals prefer to have those administrative issues successfully resolved by the project leader so that they can concentrate their efforts on the technical aspects. The essential elements of managing a group of technical professionals involve identifying the unique characteristics and needs of the group and then developing the means of satisfying those unique needs.

The nature of manufacturing and automation projects calls for the involvement of technical human resources. Recognizing the peculiar characteristics of technical professionals is one of the first steps in simplifying project management functions. Every manager must appreciate the fact that the cooperation or the lack of cooperation from technical professionals can have a significant effect on the overall management process. The success of a project can be enhanced or impeded by the management style utilized.

WORK SIMPLIFICATION

Work simplification is the systematic investigation and analysis of planned and existing work systems and methods for the purpose of developing easier, quicker, less fatiguing, and more economical ways of generating high-quality goods and services.

Work simplification facilitates the contentment of workers, which invariably leads to better performance. Consideration must be given to improving the product or service, raw materials and supplies, the sequence of operations, tools, work place, equipment, and hand and body motions. Work simplification analysis helps in defining, analyzing, and documenting work methods.

REFERENCES

Badiru, Adedeji B., "Incorporating Learning Curve Effects into Critical Resource Diagramming," *Project Management Journal*, Vol. 26, No. 2, June 1995, pp. 38–45.

Badiru, Adedeji B., "Activity-Resource Assignments Using Critical Resource Diagramming," *Project Management Journal*, Vol. 24, No. 3, September 1993, pp. 15–21.

Badiru, Adedeji B., "Critical Resource Diagram: A New Tool for Resource Management," *Industrial Engineering*, Vol. 24, No. 10, 1992, pp. 58–59, 65.

7 Techniques for Project Forecasting

Managing industrial projects effectively requires good information and good analysis. Several analytical tools are available for analyzing industrial production systems. Prior to proceeding to the project management phase of an industrial project, a good understanding of the industrial system is required. Forecasting can furnish a crucial insight into the intrinsic characteristics of the project. This chapter presents basic techniques of forecasting that are suitable for application to industrial projects. Forecasting can be used for project cost and time estimation purposes.

FORECASTING TECHNIQUES

Forecasting is an important aspect of project planning and control, as it provides the information needed to make good project decisions. Some of the basic techniques for forecasting include regression analysis, time series analysis, computer simulation, and neural networks. There are two basic types of forecasting: *intrinsic forecasting* and *extrinsic forecasting*. Intrinsic forecasting is based on the assumption that historical data can adequately describe the problem scenario to be forecasted. With intrinsic forecasting, forecasting models based on historical data use extrapolation to generate estimates for the future. Intrinsic forecasting involves the following steps:

- Collecting historical data
- Developing a quantitative forecasting model based on the data collected
- Generating forecasts recursively for the future
- Revising the forecasts as new data elements become available

Extrinsic forecasting looks outward to external factors and assumes that internal forecasts can be correlated to external factors. For example, an internal forecast of the demand for a new product may be based on external forecasts of household incomes. Good forecasts are predicated on the availability of good data. Consequently, reliable project data are essential for project forecasting.

FORECASTING BASED ON AVERAGES

The most common forecasting techniques are based on averages. Sophisticated quantitative forecasting models can be formulated from basic average formulas. The traditional techniques of forecasting based on averages are presented in the following paragraphs.

Simple Average Forecast. In this method, the forecast for the next period is computed as the arithmetic average of the preceding data points. This is often referred to as "average to date." That is:

$$f_{n+1} = \frac{\sum_{t=1}^{n} d_t}{n},$$

where

f_{n+1} = forecast for period $n + 1$
d = data for period
n = number of preceding periods for which data are available.

Period Moving-Average Forecast. In this method, the forecast for the next period is based only on the most recent data values. Each time a new value is included, the oldest value is dropped. Thus, the average is always computed from a fixed number of values. This is represented as follows:

$$f_{n+1} = \frac{\sum_{t=n-T+1}^{n} d_t}{T}$$
$$= \frac{d_{n-T+1} + d_{n-T+2} + ... + d_{n-1} + d_n}{T},$$

where

f_{n+1} = forecast for period $n + 1$
d_t = datum for period t
T = number of preceding periods included in the moving-average calculation,
n = current period at which forecast of f_{n+1} is calculated.

The moving-average technique is an after-the-fact approach. As T data points are needed to generate a forecast, we cannot generate forecasts for the first $T-1$ periods. But this shortcoming is quickly overcome as more data points become available.

Weighted-Average Forecast. The weighted-average forecast method is based on the assumption that some data points may be more significant than others in generating future forecasts. For example, the most recent data points may have greater weight than very old data points in the calculation of future estimates. This is expressed as:

$$f_{n+1} = \frac{\sum_{t=1}^{n} w_t d_t}{\sum_{t=1}^{n} w_t}$$
$$= \frac{w_1 d_1 + w_2 d_2 + ... + w_n d_n}{w_1 + w_2 + ... + w_n},$$

where

f_{n+1} = weighted-average forecast for period $n+1$

d_t = datum for period t
T = number of preceding periods included in the moving-average calculation,
n = current period at which forecast of f_{n+1} is calculated
w_t = weight of data point t.
The w_ts are the respective weights of the data points such that

$$\sum_{t=1}^{n} w_t = 1.0$$

Weighted T-Period Moving-Average Forecast. In this technique, the forecast for the next period is computed as the weighted average of past data points over the last T time periods. That is

$$f_{n+1} = w_1 d_n + w_2 d_{n-1} + \ldots + w_T d_{n-T+1},$$

where w_is are the respective weights of the data points such that the weights sum up to 1.0. That is,

$$\Sigma w_i = 1.0$$

Exponential Smoothing Forecast. This is a special case of the weighted moving-average forecast. The forecast for the next period is computed as the weighted average of the immediate past data point and the forecast of the previous period. In other words, the previous forecast is adjusted based on the deviation (forecast error) of that forecast from the actual data. That is,

$$f_{n+1} = \alpha d_n + (1 - \alpha) f_n$$
$$= f_n + \alpha (d_n - f_n),$$

where
$f_n + 1$ = exponentially weighted-average forecast for period $n + 1$
d_n = datum for period n
f_n = forecast for period n
α = smoothing factor (real number between 0 and 1).
    A low smoothing factor gives a high degree of smoothing, while a high value moves the forecast closer to actual data. The trade-off is in whether or not a smooth predictive curve is desired versus a close echo of actual data.

REGRESSION ANALYSIS

Regression analysis is a mathematical procedure for attributing the variability of one quantity to the changes in one or more other variables. It is sometimes called line fitting or curve fitting. Regression analysis is an important statistical tool that can be applied to many prediction and forecasting problems for project management purposes. The primary function of regression analysis is to develop a model that expresses the relationship between a dependent variable and one or more independent variables.

The effectiveness of a regression model is often tested by analysis of variance (ANOVA), which is a technique for breaking down the variance in a statistical sample into components that can be attributed to each factor affecting that sample. One major purpose of ANOVA is model testing. Model testing is important because of the serious consequences of erroneously concluding that a regression model is good when, in fact, it has little or no significance to the data. Model inadequacy often implies an error in the assumed relationships between the variables, poor data, or both. A validated regression model can be used for the following purposes:

1. Prediction/forecasting
2. Description
3. Control

Regression Relationships. Sometimes, the desired result from a regression analysis is an equation describing the best fit to the data under investigation. The "least squares" line drawn through the data is the line of best fit. This line may be linear or curvilinear depending on the dispersion of the data. The linear situation exists in those cases where the slope of the regression equation is a constant. The nonconstant slope indicates a curvilinear relationship. A plot of the data, called a scatter plot, will usually indicate whether a linear or nonlinear model will be appropriate. The major problem with the nonlinear relationship is the necessity of assuming a functional relationship before accurately developing the model. Example of regression models (simple linear, multiple, and nonlinear) are presented here:

$$Y = \beta_0 + \beta_1 x + \varepsilon$$
$$Y = \beta_0 + \beta_1 x_1 + \beta_2 x_2 + \varepsilon$$
$$Y = \beta_0 + \beta_1 x_1^{\alpha 1} + \beta_2 x_2^{\alpha 2} + \varepsilon$$
$$Y = \beta_0 + \beta_1 x_1^{\alpha 1} + \beta_2 x_2^{\alpha 2} + \beta_{12} x_1^{\alpha 3} x_2^{\alpha 4} + \varepsilon,$$

where Y is the dependent variable, the x_is are the independent variables, the β_is are the model parameters, and ε is the error term. The error terms are assumed to be independent and identically distributed normal random variables with mean of zero and variance with a magnitude of σ^2.

Prediction. A major use of regression analysis is prediction or forecasting. Prediction can be of two basic types: interpolation and extrapolation. Interpolation predicts values of the dependent variable over the range of the independent variable or variables. Extrapolation involves predictions outside the range of the independent variables. Extrapolation carries a risk in the sense that projections are made over a data range that is not included in the development of the regression model. There is some level of uncertainty about the nature of the relationships that may exist outside the study range. Interpolation can also create a problem when the values of the independent variables are widely spaced.

Control. Extreme care is needed in using regression for control. The difficulty lies in assuming a functional relationship when, in fact, none may exist. Suppose, for example, that regression shows a relationship between chemical content in a product and noise level in the process room. Suppose further that the real reason for this relationship is that the noise level increases as the machine speed increases, and that the higher machine speed produces higher chemical content. It would be erroneous to assume a functional relationship between the noise level in the room and the chemical content in the product. If this relationship does exist, then changes in the noise level could control chemical content. In this case, the real functional relationship exists between machine speed and chemical content. It is often difficult to prove functional relationships outside a laboratory environment because many extraneous and intractable factors may have an influence on the dependent variable. A simple example of the use of functional relationship for control can be seen in the following familiar electrical circuits' equation:

$$I = V/R,$$

where V is voltage, I is electrical current, and R is the resistance. The current can be controlled by changes in the voltage, the resistance, or both. This particular equation, which has been experimentally validated, can be used as a control design.

PROCEDURE FOR REGRESSION ANALYSIS

Problem Definition: Failure to properly define the scope of the problem could result in useless conclusions. Time can be saved throughout all phases of a regression study by knowing, as precisely as possible, the purpose of the required model. A proper definition of the problem will facilitate the selection of the appropriate variables to include in the study.

Selection of Variables: Two very important factors in the selection of variables are ease of data collection and expense of data collection. Ease of data collection deals with the accessibility and the desired form of data. We must first determine if the data can be collected, and if so, how difficult the process will be. In addition, the economic question is of prime importance. How expensive will the data be to collect and compile into a useable form? If the expense cannot be justified, then the variable under consideration may, by necessity, be omitted from the selection process.

Test of Significance of Regression: After the selection and compilation of all possible relevant variables, the next step is a test for the significance of the regression. The test should help avoid wasted effort on the use of an invalid model. The test for the significance of regression is to see whether at least one of the variable coefficients in the regression equation is statistically different from zero. A test indicating that none of the coefficients is significantly different from zero implies that the best approximation of the data is a straight line through the data at the average value of the dependent variable regardless of the values of the independent variables. The significance level of the data is an indication of how probable it is that one has erroneously assumed a model's validity.

Coefficient of Determination. The coefficient of multiple determination, denoted by R^2, is used to judge the effectiveness of regression models containing multiple variables (i.e., multiple regression model). It indicates the proportion of the variation in the dependent variable that is explained by the model. The coefficient of multiple determination is defined as

$$R^2 = \frac{SSR}{SST}$$
$$= 1 - \frac{SSE}{SST},$$

where
SSR = sum of squares due to the regression model
SST = sum of squares total
SSE = sum of squares due to error.
R^2 measures the proportionate reduction of total variation in the dependent variable accounted for by a specific set of independent variables. The coefficient of multiple determination, R^2, reduces to the *coefficient of simple determination, r^2,* when there is only one independent variable in the regression model. R^2 is equal to 0 when all the coefficients, b_k, in the model are 0. That is, there is no regression fit at all. R^2 is equal to 1 when all data points fall directly on the fitted response surface. Thus, we have

$$0.0 \leq R^2 \leq 1.0.$$

The following points should be noted about regression modeling:

1. A large R^2 does not necessarily imply that the fitted model is a useful one. For example, observations may have been taken at only a few levels of the independent variables. In such a case, the fitted model may not be useful because most predictions would require extrapolation outside the region of observations. For example, for only two data points, the regression line passes perfectly through the two points and the R^2 value will be 1.0. In that case, despite the high R^2, there will be no useful prediction capability.
2. Adding more independent variables to a regression model can only increase R^2 and never reduce it. This is because the error sum of squares (SSE) cannot become larger with more independent variables, and the total sum of squares (SST) is always the same for a given set of responses.
3. Regression models developed under conditions where the number of data points is roughly equal to the number of variables will yield high values of R^2 even though the model may not be useful. For example, for only two data points, the regression line will pass perfectly through the two points and R^2 will be 1.0, but the model would have no useful prediction capability.

The strategy for using R^2 to evaluate regression models should not entirely focus on maximizing the magnitude of R^2. Rather, the intent should be to find the point at which adding more independent variables is not worthwhile in terms of the overall effectiveness of the regression model. For most practical situations, R^2 values greater than 0.62 are considered acceptable. As R^2 can often be made larger by including a large number of independent variables, it is sometimes suggested that a modified measure that recognizes the number of independent variables in the model be used. This modified measure is referred to as the "adjusted coefficient of multiple determination," or R_a^2. It is defined mathematically as follows:

$$R_a^2 = 1 - \left(\frac{n-1}{n-p}\right)\frac{\text{SSE}}{\text{SST}},$$

where
    n = number of observations used to fit the model
    p = number of coefficients in the model (including the constant term)
    $p - 1$ = number of independent variables in the model.
    R_a^2 may actually become smaller when another independent variable is introduced into the model. This is because the decrease in SSE may be more than offset by the loss of a degree of freedom in the denominator, $n - p$.

The *coefficient of multiple correlation* is defined as the positive square root of R^2. That is,

$$R = \sqrt{R^2}$$

Thus, the higher the value of R^2, the higher the correlation in the fitted model will be.

Residual Analysis. A residual is the difference between the predicted value computed from the fitted model and the actual value from the data. The ith residual is defined as

$$e_i = Y_i - \hat{Y}_i,$$

where Y_i is the actual value and $\hat{Y}_i$ is the predicted value. The sum of squares of errors, SSE, and the mean square error, MSE, are computed as follows:

$$\text{SSE} = \sum_i e_i^2$$

$$\sigma^2 \approx \frac{\sum_i e_i^2}{n-2} = \text{MSE},$$

where n is the number of data points. A plot of residuals versus predicted values of the dependent variable can be very revealing. The plot for a good regression model will have a random pattern. A noticeable trend in the residual pattern indicates a problem with the model. Some possible reasons for an invalid regression model include the following:

- Insufficient data
- Important factors not included in model

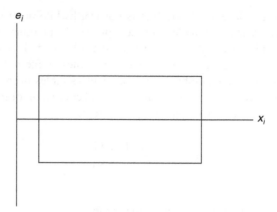

FIGURE 7.1   Ideal residual pattern.

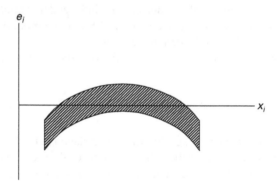

FIGURE 7.2   Residual pattern for nonlinearity.

- Inconsistency in data
- Nonexistence of any functional relationship

Graphical analysis of residuals is important for assessing the appropriateness of regression models. The different possible residual patterns are shown in Figures 7.1 through 7.6. When we plot the residuals versus the independent variable, the result should ideally appear as shown in Figure 7.1. Figure 7.2 shows a residual pattern indicating nonlinearity of the regression function. Figure 7.3 shows a pattern suggesting nonconstant variance (i.e., variation in σ^2). Figure 7.4 presents a residual pattern implying interdependence of the error terms. Figure 7.5 shows a pattern depicting the presence of outliers. Figure 7.6 represents a pattern suggesting the omission of independent variables.

TIME SERIES ANALYSIS

Time series analysis is a technique that attempts to predict the future by using historical data. The basic principle of time series analysis is that the sequence of observations is

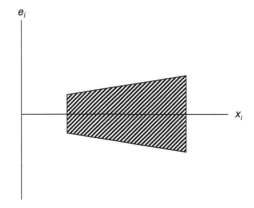

FIGURE 7.3   Residual pattern for nonconstant variance.

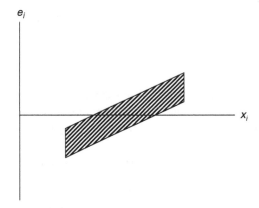

FIGURE 7.4   Residual pattern for interdependence of error terms.

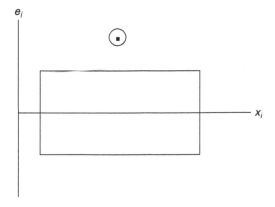

FIGURE 7.5   Residual pattern for presence of outliers.

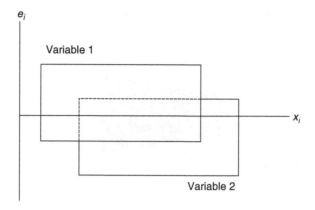

FIGURE 7.6 Residual pattern for omission of independent variables.

based on jointly distributed random variables. The time series observations denoted by $Z_1, Z_2, \ldots , Z_T$ are assumed to be drawn from some joint probability density function of the form,

$$f_1, \ldots , T\,(Z_1, \ldots , T).$$

The objective of time series analysis is to use the joint density to make probability inferences about future observations. The concept of stationarity implies that the distribution of the time series is invariant with regard to any time displacement. That is,

$$f\,(Z_t, \ldots , Z_{t+k}) = f\,(Z_{t+m}, \ldots , Z_{t+m+k}),$$

where t is any point in time and k and m are any pair of positive integers.

A stationary time series process has a constant variance and remains stable around a constant mean with respect to time reference. Thus,

$$E(Z_t) = E(Z_{t+m})$$
$$V(Z_t) = V(Z_{t+m})$$
$$Cov(Z_t, Z_{t+1}) = Cov(Z_{t+k}, Z_{t+k+1})$$

Nonstationarity in a time series may be recognized in a plot of the series. A widely scattered plot with no tendency for a particular value is an indication of nonstationarity.

Stationarity and Data Transformation. In some cases where nonstationarity exists, some form of data transformation may be used to achieve stationarity. For most time series data, the usual transformation that is employed is "differencing." Differencing involves the creation of a new series by taking differences between successive periods of the original series. For example, first, regular differences are obtained by:

$$w_t = Z_t - Z_{t-1}.$$

To develop a time series forecasting model, it is necessary to describe the relationship between a current observation and previous observations. Such relationships are described by the sample autocorrelation function defined as:

$$r_j = \frac{\sum_{t=1}^{T-j}\left(Z_t - \bar{Z}\right)\left(Z_{t+j} - \bar{Z}\right)}{\sum_{t=1}^{T}\left(Z_t - \bar{Z}\right)^2}, \quad j = 0,1, \ldots ,T-1,$$

where

T = number of observations
Z_t = observation for time t
$\bar{Z}$ = sample mean of the series
j = number of periods separating pairs of observations
r_j = sample estimate of the theoretical correlation coefficient.
The coefficient of correlation between two variables Y_1 and Y_2 is defined as

$$\rho_{12} = \frac{\sigma_{12}}{\sigma_1 \sigma_2},$$

where

σ_1 and σ_2 are the standard deviations of Y_1 and Y_2, respectively
σ_{12} is the covariance between Y_1 and Y_2.
The standard deviations are the positive square roots of the variances, defined as follows:

$$\sigma_1^2 = E\left[\left(Y_1 - \mu_1\right)^2\right]$$
$$\sigma_2^2 = E\left[\left(Y_2 - \mu_2\right)^2\right].$$

The covariance, σ_{12}, is defined as

$$\sigma_{12} = E\left[\left(Y_1 - \mu_1\right)\left(Y_2 - \mu_2\right)\right],$$

which will be zero if Y_1 and Y_2 are independent. Thus, when $\sigma_{12} = 0$, we also have $\rho_{12} = 0$. If Y_1 and Y_2 are positively related, then σ_{12} and ρ_{12} are both positive. If Y_1 and Y_2 are negatively related, then σ_{12} and ρ_{12} are both negative. The correlation coefficient is a real number between -1 and $+1$:

$$1.0 \le \rho_j \le +1.0.$$

A time series modeling procedure involves the development of a discrete linear stochastic process in which each observation, Z_t, may be expressed as

$$Z_t = \mu + \mu_t + \Psi_1 u_{t-1} + \Psi_2 u_{t-2} + \ldots$$

where μ is the mean of the process, and the Ψ_is are model parameters, which are functions of the autocorrelations. Note that this is an infinite sum indicating that the current observation at time t can be expressed in terms of all previous observations from the past. In a practical sense, some of the coefficients will be zero after some finite point q in the past. The u_ts form the sequence of independently and identically distributed random disturbances with mean zero and variance sigma sub u^2. The expected value of the series is obtained by

$$E(Z_t) = \mu + E(u_t + \Psi_1 u_{t-1} + \Psi_2 u_{t-2} + \dots + \dots + \dots)$$
$$= \mu + E(u_t)[1 + \Psi_1 + \Psi_2 + \dots]$$

Stationarity of the time series requires that the expected value be stable. That is, the infinite sum of the coefficients should be convergent to a fixed value, c, as follows:

$$\sum_{i=0}^{\infty} \Psi_i = c$$

where $\Psi_0 = 1$ and c is a constant. The theoretical variance of the process, denoted by γ_0, can be derived as:

$$\gamma_0 = E[Z_t - E(Z_t)]^2$$
$$= E[(\mu + u_t + \Psi_1 y_{t-1} + \Psi_2 u_{t-2} \dots) - \mu]^2$$
$$= E[u_t + \Psi_1 y_{t-1} + \Psi_2 u_{t-2} \dots]^2$$
$$= E[u_t^2 + \Psi_1^2 u_{t-1}^2 + \Psi_2^2 u_{t-2}^2 + \dots] + E[\text{cross} - \text{products}]$$
$$= E[u_t^2] + \Psi_1^2 E[u_{t-1}^2] + \Psi_2^2 E[u_{t-2}^2] + \dots$$
$$= \sigma_u^2 + \Psi_1^2 \sigma_u^2 + \Psi_2^2 \sigma_u^2 + \dots$$
$$= \sigma_u^2 (1 + \Psi_1^2 + \Psi_2^2 + \dots)$$
$$= \sigma_u^2 \sum_{i=0}^{\infty} \Psi_i^2,$$

where σ_u^2 represents the variance of the u_ts. The theoretical covariance between Z_t and Z_{t+j}, denoted by γ_j, can be derived in a similar manner to obtain the following:

$$\gamma_j = E\{[Z_t - E(Z_t)][Z_{t+j} - E(Z_{t+j})]\}$$
$$= \sigma_u^2 (\Psi_j + \Psi_1 \Psi_{j+1} + \Psi_2 \Psi_{j+2} + \dots)$$
$$= \sigma_u^2 \sum_{i=0}^{\infty} \Psi_i \Psi_{i+j}.$$

Sample estimates of the variances and covariances are obtained by

$$c_j = \frac{1}{T} \sum_{t=1}^{T-j} (Z_t - \overline{Z})(Z_{t+j} - \overline{Z}), \quad j = 0, 1, 2, \dots$$

The theoretical autocorrelations are obtained by dividing each of the autocovariances, γ_j, by γ_0. Thus, we have

$$\rho_j = \frac{\gamma_j}{\gamma_0}, \quad j = 0, 1, 2, \ldots,$$

and the sample autocorrelation is obtained by

$$r_j = \frac{c_j}{c_0}, \quad j = 0, 1, 2, \ldots.$$

Moving-Average Processes. If it can be assumed that $\Psi_i = 0$ for some $i > q$, where q is an integer, then our time series model can be represented as

$$z_t = \mu + u_t + \Psi_1 u_{t-1} + \Psi_2 u_{t-2} + \cdots + \cdots + \Psi_q u_{t-q},$$

which is referred to as a moving-average process of order q, usually denoted as MA(q). For notational convenience, we will denote the truncated series as presented below:

$$Z_t = \mu + u_t - \theta_1 u_{t-1} - \theta_2 u_{t-2} - \cdots - \theta_q u_{t-q},$$

where $\theta_0 = 1$. Any MA(q) process is stationary as the condition of convergence for the Ψ_is becomes

$$(1 + \Psi_1 + \Psi_2 + \cdots) = (1 - \theta_1 - \theta_2 - \cdots - \theta_q)$$
$$= 1 - \sum_{i=0}^{q} \theta_i \quad,$$

which converges since q is finite. The variance of the process now reduces to

$$\gamma_0 = \sigma_u^2 \sum_{i=0}^{q} \theta_i.$$

We now have the autocovariances and autocorrelations defined, respectively, as follows:

$$\gamma_j = \sigma_u^2 \left(-\theta_j + \theta_1 \theta_{j+1} + \cdots + \theta_{q-j} \theta_q \right), \quad j = 1, \ldots, q,$$

where $\gamma_j = 0$ for $j > q$.

$$\rho_j = \frac{\left(-\theta_j + \theta_1 \theta_{j+1} + \cdots + \theta_{q-j} \theta_q \right)}{\left(1 + \theta_1^2 + \cdots + \theta_q^2 \right)}, \quad j = 1, \ldots, q,$$

where $\rho_j = 0$ for $j > q$.

Autoregressive Processes. In the preceding section, the time series, Z_t, is expressed in terms of the current disturbance, u_t, and past disturbances, u_{t-i}. An alternative is to express Z_t, in terms of the current and past observations, Z_{t-i}. This is achieved by rewriting the time series expression as:

$$u_t = Z_t - \mu - \Psi_1 u_{t-1} - \Psi_2 u_{t-2} - \cdots$$
$$u_{t-1} = Z_{t-1} - \mu - \Psi_1 u_{t-2} - \Psi_2 u_{t-3} - \cdots$$
$$u_{t-2} = Z_{t-2} - \mu - \Psi_1 u_{t-3} - \Psi_2 u_{t-4} - \cdots$$

Successive back substitutions for the u_{t-i}s yield the following:

$$u_t = \pi_1 Z_{t-1} - \pi_2 Z_{t-2} - \cdots - \delta,$$

where π_is and δ are model parameters and are functions of Ψ_is and μ. We can then rewrite the model as

$$Z_t = \pi_1 Z_{t-1} + \pi_2 Z_{t-2} + \cdots + \pi_p Z_{t-p} + \delta + u_t$$

which is referred to as an *autoregressive process of order p*, usually denoted as AR(p). For notational convenience, we will denote the autoregressive process as follows:

$$Z_t = \phi_1 Z_{t-1} + \phi_2 Z_{t-2} + \cdots + \phi_p Z_{t-p} + \delta + u$$

Thus, AR processes are equivalent to MA processes of infinite order. Stationarity of AR processes is confirmed if the roots of the characteristic equation lie outside the unit circle in the complex plane:

$$\left(1 - \phi_1 x - \phi_2 x^2 - \cdots - \phi_p x^p\right) = 0$$

where x is a dummy algebraic symbol. If the process is stationary, then we should have:

$$
\begin{aligned}
E(Z_t) &= \phi_1 E(Z_{t-1}) + \phi_2 E(Z_{t-2}) + \cdots + \phi_p E(Z_{t-p}) + \delta + E(u_t) \\
&= \phi_1 E(Z_t) + \phi_2 E(Z_t) + \cdots + \phi_p E(Z_t) + \delta \\
&= E(Z_t)(\phi_1 + \phi_2 + \cdots + \phi_p) + \delta,
\end{aligned}
$$

which yields

$$E(Z_t) = \frac{\delta}{\left(1 - \phi_1 - \phi_2 - \cdots - \phi_p\right)}.$$

Denoting the deviation of the process from its mean by Z_t^d, the following is obtained:

$$Z_t^d = Z_t - E(Z_t) = Z_t - \frac{\delta}{\left(1 - \phi_1 - \phi_2 - \cdots - \phi_p\right)}$$

$$Z_{t-1}^d - Z_{t-1} - \frac{\delta}{\left(1 - \phi_1 - \phi_2 - \cdots - \phi_p\right)}$$

Rewriting the preceding expression yields

$$Z_{t-1} = Z_{t-1}^d + \frac{\delta}{\left(1 - \phi_1 - \phi_2 - \cdots - \phi_p\right)}$$

...

...

$$Z_{t-k} = Z_{t-k}^d + \frac{\delta}{\left(1 - \phi_1 - \phi_2 - \cdots - \phi_p\right)}$$

If we substitute the AR(p) expression into the expression for Z_t^d, we will obtain

$$Z_t^d = \phi_1 Z_{t-1} + \phi_2 Z_{t-2} + \cdots + \phi_p Z_{t-p} + \delta + u_t - \frac{\delta}{\left(1 - \phi_1 - \phi_2 - \cdots - \phi_p\right)}.$$

Successive back substitutions of Z_{t-j} into the preceding expression yields

$$Z_t^d = \phi_1 Z_{t-1}^d + \phi_2 Z_{t-2}^d + \cdots + \phi_p Z_{t-p}^d + u_t.$$

Thus, the deviation series follows the same AR process without a constant term. The tools for identifying and constructing time series models are the sample autocorrelations, r_j. For the model identification procedure, a visual assessment of the plot of r_j against j, called the sample correlogram, is used. Table 7.1 presents standard examples of *sample correlograms* and the corresponding time series models, as commonly used in industrial time series analysis.

APPLICATION OF INVENTORY MANAGEMENT TO PROJECT CONTROL

As mentioned earlier in the case of forecasting techniques, industrial projects can benefit from several analytical and quantitative decision models. Project inventory management is one quantitative approach to managing scope, cost, and schedule as a part of the project management knowledge areas. Inventoried items are an important component of any industrial project. Consequently, inventory management strategies are essential in industrial project planning and control. Tracking activities is analogous to tracking

TABLE 7.1

Identification of Some Time Series Models

Correlogram profile	Model type	Model

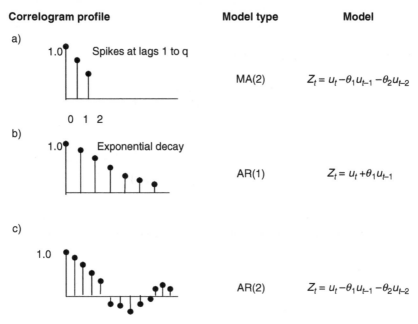

a) Spikes at lags 1 to q — MA(2) — $Z_t = u_t - \theta_1 u_{t-1} - \theta_2 u_{t-2}$

b) Exponential decay — AR(1) — $Z_t = u_t + \theta_1 u_{t-1}$

c) AR(2) — $Z_t = u_t - \theta_1 u_{t-1} - \theta_2 u_{t-2}$

Damped sine wave form

A wide variety of sample correlogram patterns can be encountered in time series analysis. It is the responsibility of the project analyst to choose an appropriate model to fit the prevailing time series data. Fortunately, several commercial statistical computer tools are available for performing time series analysis. Analysts should consult current Web postings to find the right tools for their specific application cases.

inventory items. The important aspects of inventory management for industrial project management include the following:

- Ability to satisfy work demands promptly by supplying materials from stock
- Availability of bulk rates for purchases and shipping
- Possibility of maintaining more stable and level resource or workforce

This section presents some basic inventory control models.

Economic Order Quantity Model. The economic order quantity (EOQ) model determines the optimal order quantity based on purchase cost, inventory carrying cost, demand rate, and ordering cost. The objective is to minimize the total relevant costs of inventory. For the formulation of the model, the following notations are used:

Q = replenishment order quantity (in units)
A = Fixed cost of placing an order

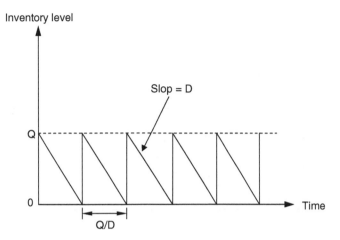

FIGURE 7.7 Basic inventory pattern.

v = variable cost per unit of the item to be inventoried
r = inventory carrying charge per dollar of inventory per unit time
D = demand rate of the item
TRC = total relevant costs per unit time.

Figure 7.7 shows the basic inventory pattern with respect to time. One complete cycle starts from a level of Q and ends at zero inventory.

The total relevant cost for order quantity Q is given by the expression below. Figure 7.8 shows the costs as functions of replenishment quantity.

$$\text{TRC}(Q) = \frac{Qvr}{2} + \frac{AD}{Q}.$$

When the TRC(Q) function is optimized with respect to Q, we obtain the expression for the EOQ as

$$\text{EOQ} = \sqrt{\frac{2AD}{vr}},$$

which represents the minimum TRCs of inventory. This formulation assumes that the cost per unit is constant regardless of the order quantity. In some cases, quantity discounts may be applicable to the inventory item. The formulation for quantity discount situation is presented in the following paragraphs.

Quantity Discount. A quantity discount may be available if the order quantity exceeds a certain level. This is referred to as the single break-point discount. The unit cost is represented as shown below. Figure 7.9 presents the price break point for a quantity discount.

$$v = \begin{cases} v_0, & 0 \le Q < Q_b \\ v_0(1-d), & Q_b \le Q \end{cases},$$

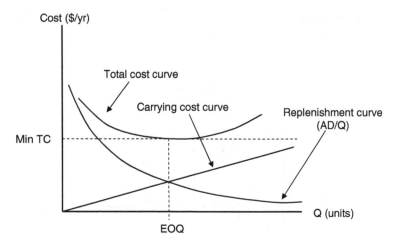

FIGURE 7.8   Inventory costs as functions of replenishment quantity.

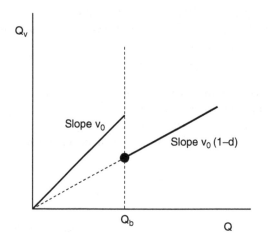

FIGURE 7.9   Price break point for quantity discount.

where
  v_0 = basic unit cost without discount
  d = discount (in decimals)
  d is applied to all units when $Q \leq Q_b$
  Q_b = break point.
  Calculation of Total Relevant Cost. For $0 \leq Q < Q_b$ we obtain

$$\text{TRC}(Q) = \left(\frac{Q}{2}\right) v_0 r + \left(\frac{A}{Q}\right) D + D v_0 .$$

For $Q_b \leq Q$ we have

$$\mathrm{TRC}(Q)_{\mathrm{discount}} = \left(\frac{Q}{2}\right) v_0 (1-d) r + \left(\frac{A}{Q}\right) D + D v_0 (1-d).$$

Note that for any given value of Q, $\mathrm{TRC}(Q)_{\mathrm{discount}} < \mathrm{TRC}(Q)$. Therefore, if the lowest point on the $\mathrm{TRC}(Q)_{\mathrm{discount}}$ curve corresponds to a value of $Q^* > Q_b$ (i.e., Q is valid), then set $Q_{\mathrm{opt}} = Q^*$.

Evaluation of the Discount Option. The trade-off between extra carrying costs and the reduction in replenishment costs should be evaluated to see if the discount option is cost justified. A reduction in replenishment costs can be achieved by two strategies:

1. Reduction in unit value
2. Fewer replenishments per unit time

Case a: If reduction in acquisition costs > extra carrying costs, then set $Q_{\mathrm{opt}} = Q_b$.

Case b: If reduction in acquisition costs < extra carrying costs, then set $Q_{\mathrm{opt}} = \mathrm{EOQ}$ with no discount.

Case c: If Q_b is relatively small, then set $Q_{\mathrm{opt}} = \mathrm{EOQ}$ with discount. The three cases are illustrated in Figure 7.10.

Based on the three cases shown in Figure 7.10, the optimal order quantity, Q_{opt}, can be found as follows:

Step 1: Compute EOQ when d is applicable:

$$\mathrm{EOQ}(\mathrm{discount}) = \sqrt{\frac{2AD}{v_0 (1-d) r}}.$$

Step 2: Compare EOQ(d) with Q_b:

If EOQ(d) $\geq Q_b$, Set $Q_{\mathrm{opt}} = \mathrm{EOQ}(d)$

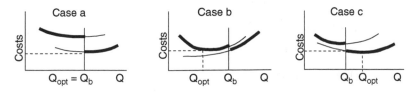

FIGURE 7.10   Cost curves for discount options.

If EOQ(d) < Q_b, go to step 3.

Step 3: Evaluate TRC for EOQ and Q_b:

$$\text{TRC}(\text{EOQ}) = \sqrt{2ADv_0 r} + Dv_0$$

$$\text{TRC}(Q_b)_{\text{dicount}} = \left(\frac{Q_b}{2}\right)v\ (1-d)r + \left(\frac{A}{Q_b}\right)D + Dv_0(1-d)$$

If $\text{TRC}(\text{EOQ}) < \text{TRC}(Q_b)$, set $Q_{\text{opt}} = \text{EOQ}(\text{no discount})$:

$$\text{EOQ}(\text{no discount}) = \sqrt{\frac{2AD}{v_0 r}}$$

If $\text{TRC}(\text{EOQ}) > \text{TRC}(Q_b)$, set $Q_{\text{opt}} = Q_b$

The following example illustrates the use of quantity discount. Suppose $d = 0.02$ and $Q_b = 100$ for the three items shown in Table 7.2.

Item 1 (Case a)
Step 1: EOQ (discount) = 19 units < 100 units
Step 2: EOQ (discount) < Q_b, go to step 3
Step 3: TRC values.

$$\text{TRC}(\text{EOQ}) = \sqrt{2(1.50)(416)(14.20)(0.24)} + 416(14.20)$$

$$= \$5972.42/\text{year}$$

$$\text{TRC}(Q_b) = \frac{100(14.20)(0.98)(0.24)}{2} + \frac{(1.50)(416)}{100} + 416(14.20)(0.98)$$

$$= \$5962.29/\text{year}$$

As TRC(EOQ) > TRC(Q_b), set Q_{opt} = 100 units.

As TRC(EOQ) > TRC(Q_b), set Q_{opt} = 100 units.

Item 2 (Case b)
Step 1: EOQ (discount) = 21 units < 100 units
Step 2: EOQ (discount) < Q_b, go to step 3
Step 3: TRC values.

$$\text{TRC}(\text{EOQ}) = \sqrt{2(1.50)(104)(3.10)(0.24)} + 104(3.10)$$

$$= \$337.64/\text{year}$$

$$\text{TRC}(Q_b) = \frac{100(3.10)(0.98)(0.24)}{2} + \frac{(1.50)(104)}{100} + 104(3.10)(0.98)$$

$$= \$353.97/\text{year}$$

$$\text{TRC}(\text{EOQ}) < \text{TRC}(Q_b), \text{ set } Q_{\text{opt}} = \text{EOQ}(\text{without discount}):$$

$$\text{EOQ} = \sqrt{\frac{2(1.50)(104)}{3.10(0.24)}}$$

$$= 20 \text{ units}$$

TABLE 7.2
Items Subject to Quantity Discount

Item	D (units/year)	v_0 ($/unit)	(A) ($)	r ($/$/yr)
Item 1	416	14.20	1.50	0.24
Item 2	104	3.10	1.50	0.24
Item 3	4160	2.40	1.50	0.24

Item 3 (Case c)
Step 1: Compute EOQ (discount)

$$\text{EOQ}(\text{discount}) = \sqrt{\frac{2(1.50)(4160)}{2.40(0.98)(0.24)}}$$

Step 2: EOQ (discount) > Q_b. Set Q_{opt} = 149 units.

Sensitivity Analysis. Sensitivity analysis involves a determination of the changes in the values of a parameter that will lead to a change in a dependent variable. It is a process for determining how wrong a decision will be if some or any of the assumptions on which the decision is based prove to be incorrect. For example, a "decision" may be dependent on the changes in the values of a particular parameter, such as inventory cost. The cost itself may in turn depend on the values of other parameters, as follows:

Subparameter ➔ Main parameter ➔ Decision

It is of interest to determine what changes in parameter values can lead to changes in a decision. With respect to inventory management, we may be interested in the cost impact of the deviation of actual order quantity from the EOQ. The sensitivity of cost to departures from EOQ is analyzed as follows.

Let p represent the level of change from EOQ:

$$|p| \le 1.0$$
$$Q' = (1-p)\text{EOQ}$$

Percentage cost penalty (PCP) is defined as follows:

$$\text{PCP} = \frac{\text{TRC}(Q') - \text{TRC}(\text{EOQ})}{\text{TRC}(\text{EOQ})}(100)$$

$$= 50\left(\frac{p^2}{1+p}\right).$$

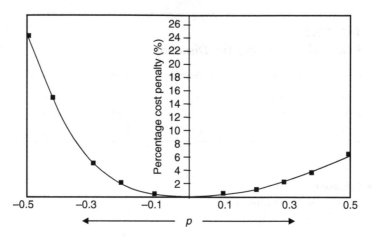

FIGURE 7.11 Sensitivity analysis based on percentage cost penalty.

A plot of the cost penalty is shown in Figure 7.11. It is seen that the cost is not very sensitive to minor departures from EOQ. We can conclude that changes within 10% of EOQ will not significantly affect the TRC. Two special inventory control algorithms are discussed in the following paragraphs.

Wagner–Whitin Algorithm. The Wagner–Whitin (W–W) algorithm is an approach to deterministic inventory modeling. It is based on dynamic programming, which is a mathematical procedure for solving sequential decision problems. The assumptions of the W–W algorithm are as follows:

- Expected demand is known for N periods into the future
- The periods are equal in length
- No stock-out or back-ordering is allowed
- All forecast demands will be met
- Ordering cost A may vary from period to period
- Orders are placed at the beginning of a period
- Order lead time is zero
- Inventory carrying cost is $(v)(r)$/unit/period
- Inventory carrying cost is charged at the beginning of a period for the units carried forward from the previous period.

Notations and Variables

In this algorithm, the following notations and variables are used:

N = number of periods in the planning horizon ($j = 1, 2, ..., N$)
t = number of periods considered when calculating first replenishment quantity ($j = 1, 2, ..., t$); where $t \leq N$
vr = inventory carrying cost/unit/period of inventory carried forward from period j to period $j+1$
A = ordering cost per order

I_j = inventory brought into period j
D_j = demand for period j
Q_j = order quantity in period j

$$\delta_j = \begin{cases} 0 \text{ if } Q_j = 0 \\ 1 \text{ if } Q_j > 1 \end{cases}$$

$$I_{j+1} = I_j + \delta_j Q_j - D_j.$$

The optimality properties of W-W algorithm are as follows:

Property 1. Replenishment takes place only when inventory level is zero.

Property 2. There is an upper limit to how far back before period j we would include D_j in a replenishment quantity.

If the requirements for period j are so large that

$$D_j(vr) > A,$$

i.e.,

$$D_j > A/vr,$$

then the optimal solution will have a replenishment at the beginning of period j. That is, inventory must go to zero at the beginning of period j. The earliest value of j where $D_j > A/vr$ is used to determine the horizon (or number of periods) to be considered when calculating the first replenishment quantity.

Let $F(t)$ = total inventory cost to satisfy demand for periods 1, 2, ... , t. So, we have:

$$\begin{aligned} F(t) = & \left[vrI_1 + A\delta_1 Q_1 \right] \\ & + \left[vr \left(I_1 + \delta_1 Q_1 - D_1 \right) + A\delta_2 Q_2 \right] \\ & + \left[vr \left(I_1 + \delta_1 Q_1 - D_1 + \delta_2 Q_2 - D_2 \right) + A\delta_3 Q_3 \right] \\ & + \cdots + \left[\cdots + A\delta_t Q_t \right] \end{aligned}$$

The expression for $F(t)$ assumes that the initial inventory is known. The objective is to minimize total inventory cost. The Applicable Propositions for the W-W Algorithm are summarized herewith:

Proposition 1: $I_j Q_j = 0$. This is because if we are planning to place an order in period j, we will not carry an inventory I_j into period j. As stock-out is not permitted, an order must be placed in period j if the inventory carried into period j is 0. Either one of I_j or Q_j must be 0.

Proposition 2: An optional policy exists such that for all periods, we have either of the following:

$$Q_j = 0$$

or

$$Q_j = \sum_{i=1}^{t} Q_i, \quad j \le t \le N.$$

This means that order quantity in period j is either zero or the total demand of some (or all) of the future periods.

Proposition 3: There exists an optimal policy such that if demand D_{j+t} is satisfied by Q_j^*, then demands $D_j, D_{j+1}, D_{j+2}, \ldots, D_{j+t-1}$ are also satisfied by Q_j^*.

Proposition 4: If an optimal inventory plan for the first t periods is given for an N-period model in which $I_j = 0$, for $j < t$, then it is possible to determine the optimal plan for periods $t+1$ through N.

Proposition 5: If an optimal inventory plan for the first t periods of an N-period model is known in which $Q_t > 0$, it is not necessary to consider periods 1 through $t-1$ in formulating the optimal plan for the rest of the periods. That is, we reinitialize the problem to start at time t and proceed forward.

Inventory analysis is needed to support procurement goals as outlined in the project management body of knowledge. Many computerized commercial systems are available for tracking inventory. Some of these can be adapted for inventory analysis in industrial project management. A good inventory management system is essential for supply chain and logistics management in an industrial project environment. This can create a competitive advantage for a company. To manage an industrial project effectively the manager or analyst must have accurate product information, which translates into accurate information about activities to be planned, scheduled, organized, and controlled.

8 Six Sigma and Lean Project Management

CONCEPT OF SIX SIGMA

The Six Sigma approach, which was originally introduced by Motorola's Government Electronics Group, has caught on quickly in the industry. Many major companies now embrace the approach as the key to high-quality industrial productivity. Six Sigma means six standard deviations from a statistical performance average. The Six Sigma approach allows for no more than 3.4 defects per million parts in manufactured goods or 3.4 mistakes per million activities in a service operation. To appreciate the effect of the Six Sigma approach, consider a process that is 99% perfect. That process will produce 10,000 defects per million parts. With Six Sigma, the process will need to be 99.99966% perfect in order to produce only 3.4 defects per million. Thus, Six Sigma is an approach that pushes the limit of perfection.

TAGUCHI LOSS FUNCTION

The philosophy of Taguchi loss function defines the concept of how deviation from an intended target creates a loss in the production process. Taguchi's idea of product quality analytically models the loss to the society from the time a product is shipped to customers. Taguchi loss function measures this conjectured loss with a quadratic function known as quality loss function (QLF), which is mathematically represented as:

$$L(y) = k(y - m)^2$$

where k is a proportionality constant, m is the target value, and y is the observed value of the quality characteristic of the product in question. The quantity $(y - m)$ represents the deviation from the target. The larger the deviation, the larger is the loss to the society. The constant k can be determined if $L(y)$, y, and m are known. Loss, in the QLF concept, can be defined to consist of several components. Examples of loss are provided in the following list:

- *Opportunity cost* of not having the service of the product because of its quality deficiency. The loss of service implies that something that should have been done to serve the society could not be done.
- *Time lost* in the search to find (or troubleshoot) the quality problem.
- *Time lost* (after finding the problem) in the attempt to solve the quality problem. The problem identification effort takes away some of the time that could have been productively used to serve the society. Thus, the society incurs a loss.

- *Productivity loss* that is incurred because of the reduced effectiveness of the product. The decreased productivity deprives the society of a certain level of service and, thereby, constitutes a loss.
- *Actual cost* of correcting the quality problem. This is, perhaps, the only direct loss that is easily recognized. But there are other subtle losses that the Taguchi method can help identify.
- *Actual loss* (e.g., loss of life) owing to a failure of the product resulting from its low quality. For example, a defective automobile tire creates a potential for traffic fatality.
- *Waste* that is generated as a result of lost time and materials because of rework and other nonproductive activities associated with low quality of work.

IDENTIFICATION AND ELIMINATION OF SOURCES OF DEFECTS

The approach uses statistical methods to find problems that cause defects. For example, the total yield (number of nondefective units) from a process is determined by a combination of the performance levels of all the steps making up the process. If a process consists of 20 steps and each step is 98% perfect, then the performance of the overall process will be:

$$(0.98)^{20} = 0.667608 \text{ (i.e., 66.7608\%)}$$

Thus, the process will produce 332,392 defects per million parts. If each step of the process is pushed to the Six Sigma limit, then the process performance will be:

$$(0.9999966)^{20} = 0.999932 \text{ (i.e., 99.9932\%)}$$

Thus, the Six Sigma process will produce only 68 defects per million parts. This is a significant improvement over the original process performance. In many cases, it is not realistic to expect to achieve the Six Sigma level of production. But the approach helps to set a quality standard and provides a mechanism for striving to reach the goal. In effect, the Six Sigma process means changing the way workers perform their tasks so as to minimize the potential for defects.

The success of Six Sigma in industry ultimately depends on the industry's ability to initiate and execute Six Sigma projects effectively. Thus, the project management approaches presented in this book are essential for realizing the benefits of Six Sigma. Project planning, organizing, team building, resource allocation, employee training, optimal scheduling, superior leadership, shared vision, and project control are all complementarily essential to implementing Six Sigma successfully. These success factors are not mutually exclusive. In many organizations, far too much focus is directed toward the statistical training for Six Sigma at the expense of proper project management development. This explains why many organizations have not been able to achieve the much-touted benefits of Six Sigma.

The success of the Toyota production system is not due to any special properties of the approach, but rather to the consistency, persistence, and dedication of Toyota organizations in building their projects around all the essential success factors. Toyota focuses on changing the organizational mindset that is required in initiating and coordinating the success factors throughout the organization. Six Sigma requires the management of multiple projects with identical mindsets throughout the organization. The success of this requirement is dependent on proper application of project management tools and techniques, as presented in the preceding chapters of this book.

ROLES AND RESPONSIBILITIES FOR SIX SIGMA

Human roles and responsibilities are crucial in executing Six Sigma projects. The different categories of team players are explained as follows:

Executive Leadership: Develops and promulgates vision, direction. Leads change and maintains accountability for organizational results (on a full-time basis).

Employee Group: Includes all employees, supports organizational vision, receives and implements Six Sigma specs, serves as points of total process improvement (TPM), exports mission statement to functional tasks, and deploys improvement practices (on full-time basis).

Six Sigma Champion: Advocates improvement projects, leads business direction, and coordinates improvement projects (on a full-time basis).

Six Sigma Project Sponsor: Develops requirements, engages project teams, leads project scoping, and identifies resource requirements (on part-time basis).

Master Belt: Trains and coaches Black Belts and Green Belts. Leads large projects, and provides leadership (on full-time basis).

Black Belt: Leads specific projects, facilitates troubleshooting, coordinates improvement groups, trains and coaches project team members (on full-time basis).

Green Belt: Participates on Black Belt teams, leads small projects (on part-time project-specific basis).

Six Sigma Project Team Members: Provides specific operational support, facilitates inward knowledge transfer, and links to functional areas (on part-time basis).

STATISTICAL TECHNIQUES FOR SIX SIGMA

Statistical process control (SPC) means controlling a process statistically. SPC originated from the efforts of the early quality control researchers. The techniques of SPC are based on basic statistical concepts normally used for statistical quality control. In a manufacturing environment, it is known that not all products are made exactly alike. There are always some inherent variations in units of the same product. The variation in the characteristics of a product provides the basis for using SPC for quality improvement With the help of statistical approaches, individual items can be studied and general inferences can be drawn about the process or batches of products from the process. As 100% inspection is difficult or impractical in many processes, SPC provides a mechanism to generalize concerning process performance. SPC uses random samples generated consecutively over time. The random samples should be representative of the general process. SPC can be accomplished through the following steps:

- Control charts ($\bar{X}$-chart and R-chart)
- Process capability analysis (nested design, C_p, and C_{pk})
- Process control (factorial design, and response surface)

CONTROL CHARTS

Two of the most commonly used control charts in industry are the $\bar{X}$ charts and the range charts (R-charts). The type of chart to be used normally depends on the kind of data collected. Data collected can be of two types: variable data and attribute data. The success of quality improvement depends on two major factors:

1. The quality of data available
2. The effectiveness of the techniques used for analyzing the data

Types of Data for Control Charts

Variable data: The control charts for variable data are

- Control charts for individual data elements (X)
- Moving-range chart (MR-chart)
- Average chart ($\bar{X}$-chart)
- Range chart (R-chart)
- Median chart
- Standard deviation chart (σ-chart)
- Cumulative sum chart (CUSUM)
Exponentially weighted moving average (EWMA)

Attribute data: The control charts for attribute data are:

- Proportion or fraction defective chart (p-chart) (subgroup sample size can vary)
- Percent defective chart (100p-chart) (subgroup sample size can vary)
- Number defective chart (np-chart) (subgroup sample size is constant)
- Number defective (c-chart) (subgroup sample size = 1)
- Defective per inspection unit (u-chart) (subgroup sample size can vary)

The statistical theory useful to generate control limits is the same for all the above charts with the exception of exponential weighted moving average (EWMA) and cumulative sum (CUSUM).

$\bar{X}$ and Range Charts

The R-chart is a time plot useful in monitoring short-term process variations, while the $\bar{X}$ chart monitors the longer term variations where the likelihood of special causes is greater over time. Both charts have control lines called upper and lower control limits, as well as the central lines. The central line and control limits are calculated from the process measurements. They are not specification limits or a percentage of the specifications, or some other arbitrary lines based on experience. Therefore, they represent what the process is capable of doing when only common cause variation exists. If only common cause variation exists, then the data will continue to fall in a random fashion within the control limits. In this case, we say the process is in a state of statistical control. However, if a special cause acts on the process, one or more data points will be outside the control limits so that the process is not in a state of statistical control.

DATA COLLECTION STRATEGIES

One strategy for data collection requires that about 20–25 subgroups be collected. Twenty to twenty-five subgroups should adequately show the location and spread of a distribution in a state of statistical control. If it happens that, owing to sampling costs or other sampling reasons associated with the process, we are unable to have 20–25 subgroups, we can still use the available samples to generate the trial control limits and update these limits as more samples are made available, because these limits will normally be wider than normal control limits and will therefore be less sensitive to changes in the process. Another approach is to use run charts to monitor the process until such time when 20–25 subgroups are made available. Then, control charts can be applied with control limits included on the charts. Other data collection strategies should consider the subgroup sample size, as well as the sampling frequency.

Subgroup Sample Size

The subgroup samples of size n should be taken as n consecutive readings from the process and not random samples. This is necessary in order to have an accurate estimate of the process common cause variation. Each subgroup should be selected from some small period of time or small region of space or product in order to assure homogeneous conditions within the subgroup. This is necessary because the variation within the subgroup is used in generating the control limits. The subgroup sample size n can be between four or five samples. This is a good size that balances the pros and cons of using large or small sample size for a control chart, as provided in the following lists.

The advantages of using small subgroup sample size are as follows:

- Estimates of process standard deviation based on the range are as good and accurate as the estimates obtained from using the standard deviation equation, which is a complex hand-calculation method.
- The probability of introducing special cause variations within a subgroup is very small.

- Range chart calculation is simple and easier to compute by hand on the shop floor by operators.

The advantages of using large subgroup sample size are as follows:

- The central limit theorem supports the fact that the process average will be more normally distributed with larger sample size.
- If the process is stable, the larger the subgroup size the better the estimates of process variability.
- A control chart based on larger subgroup sample size will be more sensitive to process changes.

The choice of a proper subgroup is very critical to the usefulness of any control chart. The following paragraphs explain the importance of subgroup characteristics:

- If we fail to incorporate all common cause variations within our subgroups, the process variation will be underestimated, leading to very tight control limits. Then the process will appear to go out of control too frequently even when there is no existence of a special cause.
- If we incorporate special causes within our subgroups, then we will fail to detect special causes as frequently as expected.

Frequency of Sampling

The problem of determining how frequently one should sample depends on several factors. These factors include, but are not limited to, the following.

- *Cost of collecting and testing samples*: The greater the cost of taking and testing samples, the less frequently we should sample.
- *Changes in process conditions*: The larger the frequency of changes to the process, the larger is the sampling frequency. For example, if process conditions tend to change every 15 min, then sample every 15 min. If conditions change every 2 hr, then sample every 2 hr.
- *Importance of quality characteristics*: The more important the quality characteristic being charted is to the customer, the more frequently the characteristic will need to be sampled.
- *Process control and capability*: The more history of process control and capability, the less frequently the process needs to be sampled.

Stable Process

A process is said to be in a state of statistical control if the distribution of measurement data from the process has the same shape, location, and spread over time. In other words, a process is stable when the effects of all special causes have been removed from a process, so that the remaining variability is only due to common causes. Figure 8.1 shows an example of a stable distribution.

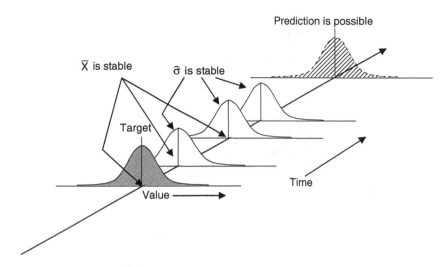

FIGURE 8.1 Stable distribution with no special causes.

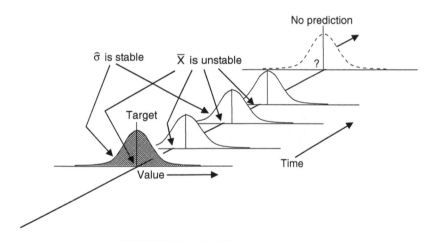

FIGURE 8.2 Unstable process average.

OUT-OF-CONTROL PATTERNS

A process is said to be unstable (*not* in a state of statistical control) if it changes from time to time because of a shifting average, or shifting variability, or a combination of shifting averages and variation. Figure 8.2 through 8.4 show examples of distributions from unstable processes.

CALCULATION OF CONTROL LIMITS

- Range (R)
  This is the difference between the highest and lowest observations:
  $R = X_{highest} - X_{lowest}$

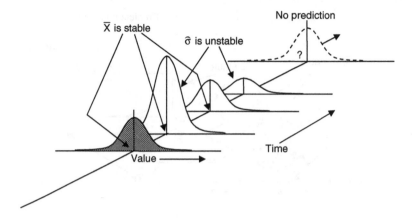

FIGURE 8.3   Unstable process variation.

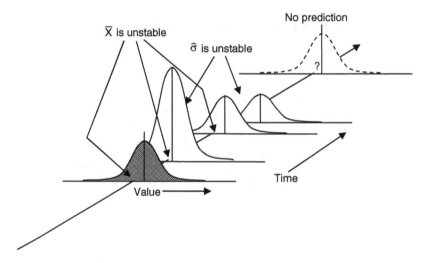

FIGURE 8.4   Unstable process average and variation.

- Center lines
  Calculate $\bar{X}$ and $\bar{R}$

$$\bar{X} = \frac{\sum X_i}{m}$$

$$\bar{R} = \frac{\sum R_i}{m}$$

where

$\bar{X}$ = overall process average
$\bar{R}$ = average range
m = total number of subgroups
n = within subgroup sample size

- Control limits based on R-chart

$$\text{UCL}_R = D_4 \bar{R}$$
$$\text{LCL}_R = D_3 \bar{R}$$

- Estimate of process variation

$$\hat{\sigma} = \frac{\bar{R}}{d_2}$$

- Control limits based on $\bar{X}$-chart
  Calculate the upper and lower control limits for the process average:

$$\text{UCL} = \bar{X} + A_2 \bar{R}$$
$$\text{LCL} = \bar{X} - A_2 \bar{R}$$

Table 8.1 shows the values of d_2, A_2, D_3, and D_4 for different values of n. These constants are used for developing variable control charts.

Plotting control charts for range and average charts:

- Plot the range chart (R-chart) first.
- If R-chart is in control, then plot $\bar{X}$ chart.
- If R-chart is not in control, identify and eliminate special causes, then delete points that are due to special causes, and recompute the control limits for the range chart. If process is in control, then plot $\bar{X}$ chart.
- Check to see if $\bar{X}$ chart is in control, if not search for special causes and eliminate them permanently.
- Remember to perform the eight trend tests.

Plotting control charts for moving range and individual control charts:

- Plot the moving range chart (MR-chart) first.
- If MR-chart is in control, then plot the individual chart (X).

TABLE 8.1
Table of Constants for Variables Control Charts

n	d_2	A_2	D_2	D_2
2	1.128	1.880	0	3.267
3	1.693	1.023	0	2.575
4	2.059	0.729	0	2.282
5	2.326	0.577	0	2.115
6	0.534	0.483	0	2.004
7	2.704	0.419	0.076	1.924
8	2.847	0.373	0.136	1.864
9	2.970	0.337	0.184	1.816
10	3.078	0.308	0.223	1.777
11	3.173	0.285	0.256	1.744
12	3.258	0.266	0.284	1.716

- If MR-chart is not in control, identify and eliminate special causes, then delete special causes points, and recompute the control limits for the moving range chart. If MR-chart is in control, then plot the individual chart.
- Check to see if individual chart is in control, if not search for special causes from out-of-control points.
- Perform the eight trend tests.

TABLE 8.2
Data for Control Chart Example

Smoothness (micro-inches)

Subgroup number	I	II	III	IV	Average	Range
1	34	33	24	28	29.75	10
2	33	33	33	29	32.00	4
3	32	31	25	28	29.00	7
4	33	28	27	36	31.00	9
5	26	34	29	29	29.50	8
6	30	31	32	28	30.25	4
7	25	30	27	29	27.75	5
8	32	28	32	29	30.25	4
9	29	29	28	28	28.50	1
10	31	31	27	29	29.50	4
11	27	36	28	29	30.00	9
12	28	27	31	31	29.25	4
13	29	31	32	29	30.25	3
14	30	31	31	34	31.50	4
15	30	33	28	31	30.50	5
16	27	28	30	29	28.50	3
17	28	30	33	26	29.25	7
18	31	32	28	26	29.25	6
19	28	28	37	27	30.00	10
20	30	29	34	26	29.75	8
21	28	32	30	24	28.50	8
22	29	28	28	29	28.50	1
23	27	35	30	30	30.50	8
24	31	27	28	29	28.75	4
25	32	36	26	35	32.25	10
26	27	31	28	29	28.75	4
27	27	29	24	28	27.00	5
28	28	25	26	28	26.75	3
29	25	25	32	27	27.25	7
30	31	25	24	28	27.00	7
Total					881.00	172

Case Example: Plotting of Control Chart

An industrial engineer in a manufacturing company was trying to study a machining process for producing a smooth surface on a torque converter clutch. The quality characteristic of interest was the surface smoothness of the clutch. The engineer then collected four clutches every hour for 30 hr and recorded the smoothness measurements in μin. Acceptable values of smoothness lies between 0 (perfectly smooth) and 45 micro-inches. The data collected by the engineer are provided in Table 8.2. Histograms of the individual and average measurements are presented in Figure 8.5.

The two histograms in Figure 8.5 show that the hourly smoothness average ranges from 27 to 32 micro-inches, much narrower than the histogram of hourly individual smoothness, which ranges from 24 to 37 micro-inches. This is due to the fact that averages have less variability than individual measurements. Therefore, whenever we plot subgroup averages on an $\bar{X}$ chart, there will always exist some individual measurements that will plot outside the control limits of an $\bar{X}$ chart. The dot plots of the surface smoothness for individual and average measurements are shown in Figure 8.6.

The descriptive statistics for individual smoothness are presented in the following table:

N	MEAN	MEDIAN	Total Range MEAN	Standard Deviation	Standard Expected MEAN
120	29.367	29.00	29.287	2.822	0.258
MIN	MAX	Q1	Q3		
24.00	37.00	28.00	31.00		

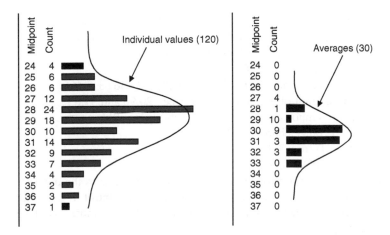

FIGURE 8.5 Histograms of individual measurements and averages for clutch smoothness.

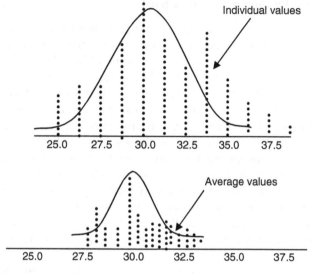

FIGURE 8.6   Dot Plots of individual measurements and averages for clutch smoothness.

The descriptive statistics for average smoothness are presented in the following table:

N	MEAN	MEDIAN	Total Range MEAN	Standard Deviation	Standard Expected MEAN
30	29.367	29.375	29.246	1.409	0.257
MIN	MAX	Q1	Q3		
26.75	32.25	28.50	30.25		

Calculations

1. Natural limit of the process $= \bar{X} \pm 3s$ (based on empirical rule).
   s = estimated standard deviation of all individual samples
   Standard deviation (special and common) $s = 2.822$
   Process average $\bar{X} = 29.367$
   Natural process limit $= 29.367 \pm 3\,(2.822) = 29.367 \pm 8.466$
   The natural limit of the process is between 20.90 and 37.83
2. Inherent (common cause) process variability $\hat{\sigma} = \bar{R}/d_2$
   $\bar{R}$ from the range chart $= 5.83$
   d_2 (for $n = 4$) from Table 8.1 $= 2.059$
   $\hat{\sigma} = \bar{R}/d_2 = 5.83/2.059 = 2.83$
   Thus, the total process variation, s, is about the same as the inherent process variability. This is because the process is in control. If the process is out of control, the total standard deviation of all the numbers will be larger than $\bar{R}/d_2$.
3. Control limits for the range chart
   Obtain constants D_3, D_4 from Table 8.1 for $n = 4$.

$D_3 = 0$

$D_4 = 2.282$

$\bar{R} = 172/30 = 5.73$

$UCL = D_4 \times \bar{R} = 2.282(5.73) = 16.16$

$LCL = D_3 \times \bar{R} = 0(5.73) = 0.0$

4. Control limits for the averages

Obtain constants A_2 from Table 8.1 for $n = 4$.

$A_2 = 0.729$

$UCL = \bar{X} + A_2(\bar{R}) = 29.367 + 0.729(5.73) = 33.54$

$LCL = \bar{X} - A_2(\bar{R}) = 29.367 - 0.729(5.73) = 25.19$

5. Natural limit of the process = $\bar{X} \pm 3(\bar{R})/d_2 = 29.367 \pm 3(2.83) = 29.367 \pm 8.49$

The natural limit of the process is between 20.88 and 37.86, which is slightly different from $\pm 3s$ calculated earlier based on the empirical rule. This is due to the fact that $\bar{R}/d_2$ is used rather than the standard deviation of all the values. Again, if the process is out of control, the standard deviation of all the values will be greater than $\bar{R}/d_2$. The correct procedure is always to use $\bar{R}/d_2$ from a process that is in statistical control.

6. Comparison with specification

As the specifications for the clutch surface smoothness is between 0 (perfectly smooth) and 45 micro-inches, and the natural limit of the process is between 20.88 and 37.86, then the process is capable of producing within the spec limits. Figure 8.7 presents the R- and $\bar{X}$ charts for clutch smoothness.

For this case example, the industrial engineer examined the charts in Figure 8.6 and concluded that the process is in a state of statistical control.

Process Improvement Opportunities

The industrial engineer realizes that if the smoothness of the clutch can be held below 15 micro-inches, then the clutch performance can be significantly improved. In this situation, the engineer can select key control factors to study in a two-level factorial or fractional factorial design.

Trend Analysis

After a process is recognized to be out of control, zone control charting technique is a logical approach to searching for the sources of the variation problems. The following eight tests can be performed using MINITAB software or other statistical software tools. For this approach, the chart is divided into three zones. Zone A is between $\pm 3\sigma$, zone B is between $\pm 2\sigma$, and zone C is between $\pm 1\sigma$.

Test 1

Pattern: One or more points falling outside the control limits on either side of the average. This is shown in Figure 8.8.

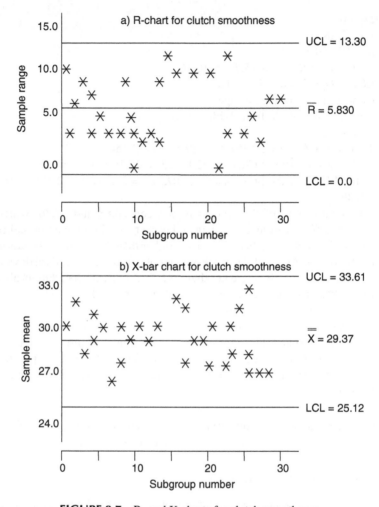

FIGURE 8.7 R- and X-charts for clutch smoothness.

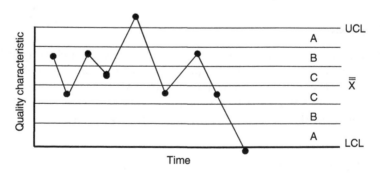

FIGURE 8.8 Test 1 for trend analysis.

Problem source: A sporadic change in the process due to special causes such as:

- Equipment breakdown
- New operator
- Drastic change in raw material quality
- Change in method, machine, or process setting

Check: Go back and look at what might have been done differently before the out-of-control point signals.

Test 2
Pattern: A run of nine points on one side of the average (Figure 8.9).

Problem source: This may be due to a small change in the level of process average. This change may be permanent at the new level.

Check: Go back to the beginning of the run and determine what was done differently at that time or prior to that time.

Test 3
Pattern: A trend of six points in a row either increasing or decreasing as shown in Figure 8.10.

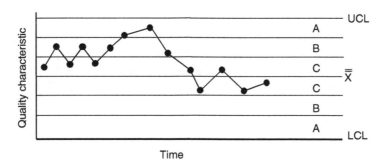

FIGURE 8.9   Test 2 for trend analysis.

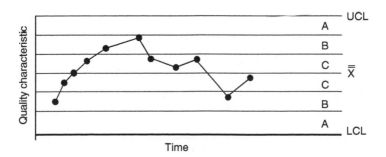

FIGURE 8.10   Test 3 for trend analysis.

Problem source: This may be due to the following:

- Gradual tool wear
- Change in characteristic such as gradual deterioration in the mixing or concentration of a chemical
- Deterioration of plating or etching solution in electronics or chemical industries

Check: Go back to the beginning of the run and search for the source of the run.

These three tests are useful in providing good control of a process. However, in addition to the above three tests, some advanced tests for detecting out-of-control patterns can also be used. These tests are based on the zone control chart.

Test 4

Pattern: Fourteen points in a row alternating up and down within or outside the control limits as shown in Figure 8.11.

Problem source: This can be due to sampling variation from two different sources such as sampling systematically from high and low temperatures, or batches with two different averages. This pattern can also occur if adjustment is being made all the time (over control).

Check: Look for cycles in the process, such as humidity or temperature cycles, or operator over control of process.

Test 5

Pattern: Two out of three points in a row on one side of the average in zone A or beyond. An example of this is presented in Figure 8.12.

Problem source: This can be due to a large, dramatic shift in the process level. This test sometimes provides early warning, particularly if the special cause is not as sporadic as in the case of Test 1.

Check: Go back one or more points in time and determine what might have caused the large shift in the level of the process.

Test 6

Pattern: Four out of five points in a row on one side of the average in zone B or beyond, as depicted in Figure 8.13.

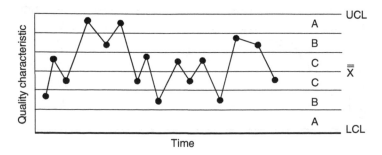

FIGURE 8.11   Test 4 for trend analysis.

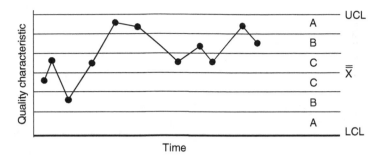

FIGURE 8.12   Test 5 for trend analysis.

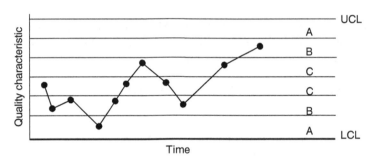

FIGURE 8.13   Test 6 for trend analysis.

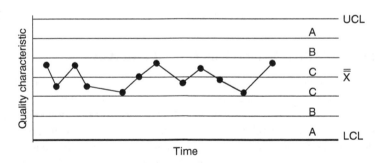

FIGURE 8.14   Test 7 for trend analysis.

Problem source: This may be due to a moderate shift in the process.
Check: Go back three or four points in time.

Test 7

Pattern: Fifteen points in a row on either side of the average in zone C as shown in Figure 8.14.

Problem source: This is due to the following:

- Unnatural small fluctuations or absence of points near the control limits.
- At first glance may appear to be a good situation, but this is not a good control.

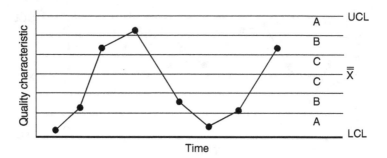

FIGURE 8.15   Test for trend analysis.

- Incorrect selection of subgroups. May be sampling from various subpopulations and combining them into a single subgroup for charting.
- Incorrect calculation of control limits.

Check: Look very close to the beginning of the pattern.

Test 8

Pattern: Eight points in a row on both sides of the center line with none in zone C. An example is shown in Figure 8.15.

Problem source: No sufficient resolution on the measurement system (see Section on measurement system).

Check: Look at the range chart and see if it is in control.

PROCESS CAPABILITY ANALYSIS FOR SIX SIGMA

Industrial process capability analysis is an important aspect of managing industrial projects. The capability of a process is the spread that contains almost all values of the process distribution. It is very important to note that capability is defined in terms of a distribution. Therefore, capability can only be defined for a process that is stable (has distribution) with common cause variation (inherent variability). It cannot be defined for an out-of-control process (which has no distribution) with variation special to specific causes (total variability). Figure 8.16 shows a process capability distribution.

CAPABLE PROCESS (CP)

A process is capable ($C_p \geq 1$) if its natural tolerance lies within the engineering tolerance or specifications. The measure of process capability of a stable process is $6\hat{\sigma}$, where $\hat{\sigma}$ is the inherent process variability is estimated from the process. A minimum value of $C_p = 1.33$ is generally used for an ongoing process. This ensures a very low reject rate of 0.007%, and therefore is an effective strategy for prevention of nonconforming items. C_p is defined mathematically as:

$$C_p = \frac{USL - LS}{6\hat{\sigma}}$$

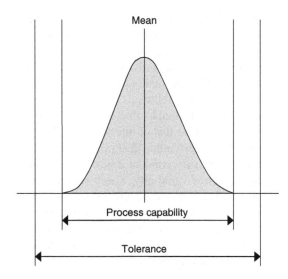

FIGURE 8.16 Process capability distribution.

$$= \frac{\text{allowable process spread}}{\text{actual process spread}}$$

where
   USL = upper specification limit
   LSL = lower specification limit

C_p measures the effect of the inherent variability only. The analyst should use $\bar{R}/d_2$ to estimate $\hat{\sigma}$ from an R-chart that is in a state of statistical control, where $\bar{R}$ is the average of the subgroup ranges and d_2 is a normalizing factor that is tabulated for different subgroup sizes (n).

We do not have to verify control before performing a capability study. We can perform the study, and then verify control after the study with the use of control charts. If the process is in control during the study, then our estimates of capabilities are correct and valid. However, if the process was not in control, we would have gained useful information, as well as proper insights as to the corrective actions to pursue.

CAPABILITY INDEX (CPK)

Process centering can be assessed when a two-sided specification is available. If the capability index (C_{pk}) is equal to or greater than 1.33, then the process may be adequately centered. C_{pk} can also be employed when there is only one-sided specification. For a two-sided specification, it can be mathematically defined as:

$$C_{pk} = \text{Minimum} \left(\frac{\text{USL} - \bar{X}}{3\hat{\sigma}}, \frac{\bar{X} - \text{LSL}}{3\hat{\sigma}} \right)$$

where $\bar{X}$ = Overall process average.

However, for a one-sided specification, the actual C_{pk} obtained is reported. This can be used to determine the percentage of observations out of specification. The overall long-term objective is to make C_p and C_{pk} as large as possible by continuously improving or reducing process variability, $\hat{\sigma}$, for every iteration so that a greater percentage of the product is near the key quality characteristic's target value. The ideal is to center the process with zero variability.

If a process is centered but not capable, one or several courses of action may be necessary. One of the actions may be that of integrating designed experiment to gain additional knowledge on the process and in designing control strategies. If excessive variability is demonstrated, one may conduct a nested design with the objective of estimating the various sources of variability. These sources of variability can then be evaluated to determine what strategies to use in order to reduce or permanently eliminate them. Another action may be that of changing the specifications or continuing production and then sorting the items. Three characteristics of a process can be observed with respect to capability, as summarized in the following text. Figure 8.17 through 8.19 present the alternate characteristics.

1. The process may be centered and capable.
2. The process may be capable but not centered.
3. The process may be centered but not capable.

Process Capability Example

Step 1: Using data for the specific process, determine if the process is capable. Let us assume that the analyst has determined that the process is in a state of statistical control. For this example, the specification limits are set at 0 (lower limit) and 45 (upper limit). The inherent process variability as determined from the control chart is:

$$\hat{\sigma} = \bar{R}/d_2 = 5.83/2.059 = 2.83.$$

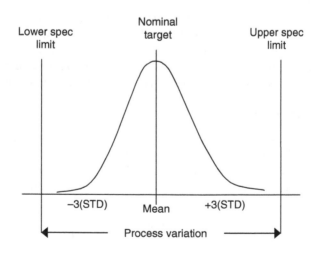

FIGURE 8.17 A process that is centered and capable.

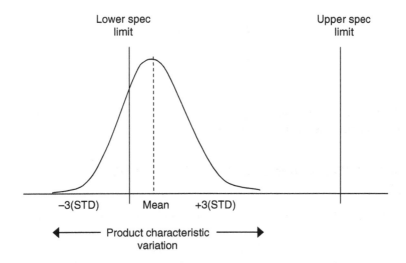

FIGURE 8.18 A process that is capable but not centered.

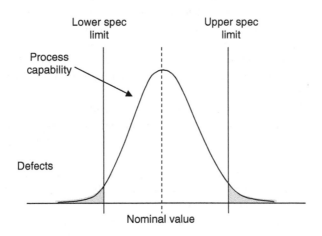

FIGURE 8.19 A process that is centered but not capable.

The capability of this process to produce within the specifications can be determined as:

$$C_p = \frac{USL - LSL}{6\hat{\sigma}} = \frac{45 - 0}{6(2.83)} = 2.650.$$

The capability of the process, $C_p = 2.65 > 1.0$ indicating that the process is capable of producing clutches that will meet the specifications of between 0 and 45. The process average is 29.367.

Step 2: Determine if the process can be adequately centered. C_{pk} = minimum (C_l and C_u) can be used to determine if a process can be centered.

$$C_u = \frac{USL - \bar{X}}{3\hat{\sigma}} = \frac{45 - 29.367}{3(2.83)} = 1.84$$

$$C_l = \frac{\bar{X} - LSL}{3\hat{\sigma}} = \frac{29.367 - 0}{3(2.83)} = 3.46$$

Therefore, the capability index, C_{pk}, for this process is 1.84. As $C_{pk} = 1.84$ is greater than 1.33, the process can be adequately centered.

POSSIBLE APPLICATIONS OF PROCESS CAPABILITY INDEX

The potential applications of process capability index are summarized here:

- *Communication:* C_p and C_{pk} have been used in industry to establish a dimensionless common language useful for assessing the performance of production processes. Engineering, quality, manufacturing, and so forth can communicate and understand processes with high capabilities.
- *Continuous improvement:* The indices can be used to monitor continuous improvement by observing the changes in the distribution of process capabilities. For example, if there were 20% of processes with capabilities between 1 and 1.67 in a month, and some of these improved to between 1.33 and 2.0 the next month, then this is an indication that improvement has occurred.
- *Audits:* There are so many various kinds of audits in use today to assess the performance of quality systems. A comparison of in-process capabilities with capabilities determined from audits can help establish problem areas.
- *Prioritization of improvement:* A complete printout of all processes with unacceptable C_p or C_{pk} values can be extremely powerful in establishing the priority for process improvements.
- *Prevention of nonconforming product:* For process qualification, it is reasonable to establish a benchmark capability of $C_{pk} = 1.33$, which will make nonconforming products unlikely in most cases.

POTENTIAL ABUSE OF C_p AND C_{pk}

In spite of its several possible applications, process capability index has some potential sources of abuse as summarized in the following list:

- *Problems and Drawbacks*: C_{pk} can increase without process improvement even though repeated testing reduces test variability; The wider the specifications, the larger is the C_p or C_{pk}, but the action does not improve the process.
- Analysts tend to focus on number rather than on process.

- *Process Control*: Analysts tend to determine process capability before statistical control has been established. Most people are not aware that capability determination is based on process common cause variation and what can be expected in the future. The presence of special causes of variation makes prediction impossible and capability index unclear.
- *Non-normality*: Some processes result in non-normal distribution for some characteristics. As capability indices are very sensitive to departures from normality, data transformation may be used to achieve approximate normality.
- *Computation*: Most computer-based tools do not use $\bar{R}/d_2$ to calculate σ.

When analytical and statistical tools are coupled with sound managerial approaches, an organization can benefit from a robust implementation of improvement strategies. One approach that has emerged as a sound managerial principle is "Lean," which has been successfully applied to many industrial operations.

LEAN PRINCIPLES AND APPLICATIONS

What is "Lean"? Lean means the identification and elimination of sources of *waste* in operations. Recall that Six Sigma involves the identification and elimination of source of *defects*. When Lean and Six Sigma are coupled, an organization can derive the double benefit of reducing waste and defects in operations; which leads to what is known as Lean-Six-Sigma. Consequently, the organization can achieve higher product quality, better employee morale, better satisfaction of customer requirements, and more effective utilization of limited resources. The basic principle of Lean is to take a close look at the elemental compositions of a process so that non-value-adding elements can be located and eliminated.

KAIZEN OF A PROCESS

By applying the Japanese concept of "*Kaizen*," which means "take apart and make better," an organization can redesign its processes to be lean and devoid of excesses. In a mechanical design sense, this can be likened to finite element analysis, which identifies how the component parts of a mechanical system fit together. It is by identifying these basic elements that improvement opportunities can be easily and quickly recognized. It should be recalled that the process of work breakdown structure in project management facilitates the identification of task-level components of an endeavor. Consequently, using a project management approach facilitates the achievement of the objectives of Lean. Figure 8.20 shows a process decomposition hierarchy that may help to identify elemental characteristic that may harbor waste.

LEAN BEYOND FADDISM

Fads come and go in industry. Over the years, we have witnessed the introduction and demise of many techniques that were hailed as the panacea of industry's ailments. Some of the techniques have survived the test of time because they do, indeed,

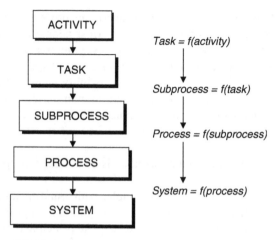

FIGURE 8.20   Hierarchy of process components.

hold some promise. Lean techniques appear to hold such promise, if it is viewed as an open-ended but focused application of the many improvement tools that have emerged over the years.

Case Example of Lean: AFSO21

The adoption of lean principles by the US Air Force has given more credence to its application. The US Air Force embarked on a massive endeavor to achieve wide-spread improvement in operational processes throughout the Air Force. The endeavor is called AFSO21 (Air Force Smart Operations for the 21st Century, or simply Air Force Smart Ops 21). This endeavor requires the implementation of appropriate project management practices at all levels. AFSO21 is a coordinated effort at achieving operational improvement in US Air Force operations throughout the rank and file of the large organization. It is an integrative process of using Lean Principles, Theory of Constraints, Six Sigma, Business Process Reengineering (BPI), Management By Objective (MBO), Total Quality Management (TQM), 6s, Project Management, and other classical management tools. However, the implementation of Lean principles constitutes about 80% of AFSO21 efforts.

As a part of tools for Lean practices and procedures, the following section presents the Lean task value rating system, which helps to compare and rank elements of a process for retention, rescoping, scaling, or elimination purposes.

LEAN TASK VALUE RATING SYSTEM

In order to identify value-adding elements of a Lean project, the component tasks must be ranked and comparatively assessed. The following method applies relative ratings to tasks. It is based on the distribution of a total point system. The total points available to the composite process or project are allocated across individual tasks. The steps are explained as follows:

1. Let T be the total points available to tasks.
2. $T = 100(n)$, where n = number of raters on the rating team.

3. Rate the value of each task on the basis of specified output (or quality) criteria on a scale of 0–100.
4. Let x_{ij} be the rating for task i by rater j.
5. Let m = number of tasks to be rated.
6. Organize the ratings by rater j as shown below:

Rating for task 1:	x_{ij}
Rating for task 2:	x_{2j}
.	.
.	.
.	.
Rating for task m:	x_{mj}
Total rating points	100

7. Tabulate the ratings by the raters as shown in Table 8.3 and calculate the overall weighted score for each Task i from the following expression:

$$w_i = \frac{1}{n}\sum_{j=1}^{n} x_{ij}$$

The w_i are used to rank order the tasks to determine the relative value-added contributions of each. Subsequently, using a preferred cutoff margin, the low or noncontributing activities can be slated for elimination.

In terms of activity prioritization, a comprehensive Lean analysis can identify the important versus unimportant and urgent versus not urgent tasks. It is within the unimportant and not urgent quadrant that one will find "waste" task elements that should be eliminated. Table 8.4 presents an example of task elements within a 20% waste elimination zone, determined using the familiar Pareto distribution format.

It is conjectured that activities that fall in the "not important" and "not urgent" zone run the risk of generating points of waste in any productive undertaking. That zone should be the first target of review for tasks that can be eliminated. Granted that there may be some "sacred cow" activities that an organization must retain for political, cultural, or regulatory reasons, but attempts should still be made to categorize all task elements of a project. The long-established industrial engineering principle of

---

TABLE 8.3
Lean Task Rating Matrix

	Rating by rater j = 1	Rating by rater j = 2	...	...	Rating by rater n	Total points for task i	wi
Rating for task $i = 1$							
Rating for task $i = 2$							
...							
...							
Rating for task m							
Total points from rater j	100	100	...	...	100	100n	

TABLE 8.4
Pareto Analysis of Unimportant
Process Task Elements

	Urgent	Not urgent
Important	20%	80%
Not important	80%	20%

---

time-and-motion studies is making a come back because of the increased interest in eliminating waste in Lean initiatives.

COMBINING LEAN, SIX SIGMA, AND PROJECT MANAGEMENT

As can be seen in the details presented in this book, both Lean and Six Sigma use analytical tools as the basis for pursuing their goals. But the achievement of those goals is predicated on having a structured approach to the activities of production. If proper project management is practiced at the outset on an industrial endeavor, it will pave the way for achieving Six Sigma results and realizing Lean outcomes. The key in any project endeavor is to have a structured design of the project so that diagnostic and corrective steps can easily be pursued. If the proverbial "garbage" is allowed to creep into a project, it would take much more time, effort, and cost to achieve a Lean-Six-Sigma cleanup.

9 Project Risk Analysis

Uncertainty is a reality in any decision-making process. Project analysts must take risk and uncertainty into consideration when dealing with industrial projects. An industrial project analyst should be able to identify, analyze, quantify, evaluate, track, and develop mitigation strategies for project risks.

FACTORS INFLUENCING DECISIONS

Traditional decision theory classifies decisions under three different influences:

1. *Decision under Certainty*: Made when possible events or outcomes of a decision can be positively determined.
2. *Decisions under Risk*: Made using information on the probability that a possible event or outcome will occur.
3. *Decisions under Uncertainty*: Made by evaluating possible events or outcomes without information on the probability that the events or outcomes will occur.

Many authors make a distinction between decisions under risk and under uncertainty. In the literature, decisions made under risk, as defined in the preceding list, is being increasingly incorporated into those made under uncertainty. In this book, no special distinction will be made between risk and uncertainty. Some of the chapters in this book contain a number of procedures to illustrate how project decisions may be made under uncertainty. Some of the parameters that normally change during a project life cycle include project costs, time requirements, and performance specifications. The uncertainties associated with these parameters are a concern for project managers. Cost, time, and performance must be managed throughout the project life cycle.

COST UNCERTAINTIES

In an inflationary economy, project costs can become very dynamic and intractable. Cost estimates include various tangible and intangible components of a project, such as machines, inventory, training, raw materials, design, and personnel wages. Costs can change during a project for a number of reasons including:

- External inflationary trends
- Internal cost adjustment procedures
- Modification of work process
- Design adjustments

195

- Changes in cost of raw materials
- Changes in labor costs
- Adjustment of work breakdown structure
- Cash flow limitations
- Effects of tax obligations

These cost changes and others combine to create uncertainties in the project's cost. Even when the cost of some of the parameters can be accurately estimated, the overall project cost may still be uncertain because of the few parameters that cannot be accurately estimated.

SCHEDULE UNCERTAINTIES

Unexpected engineering change orders (ECO) and other changes in a project environment may necessitate schedule changes, which introduce uncertainties to the project. The following are some of the reasons project schedules change:

- Task adjustments
- Changes in scope of work
- Changes in delivery arrangements
- Changes in project specification
- Introduction of new technology

PERFORMANCE UNCERTAINTIES

Performance measurement involves observing the value of parameters during a project, and comparing the actual performance, based on the observed parameters, with the expected performance. Performance control then takes appropriate actions to minimize the deviations between actual performance and expected performance. Project plans are based on the expected performance of the project parameters. Performance uncertainties exist when expected performance cannot be defined in definite terms. As a result, project plans require a frequent review.

The project management team must have a good understanding of the factors that can have a negative impact on the expected project performance. If at least some of the sources of deficient performance can be controlled, then the detrimental effects of uncertainties can be alleviated. The most common factors that can influence project performance include the following:

- Redefinition of project priorities
- Changes in management control
- Changes in resource availability
- Changes in work ethics
- Changes in organizational policies and procedures
- Changes in personnel productivity
- Changes in quality standards

To minimize the effect of uncertainties in project management, a good control must be maintained over the various sources of uncertainty discussed earlier. The same analytic tools that are effective for one category of uncertainties should also work for other categories.

DECISION TABLES AND TREES

Decision tree analysis is used to evaluate sequential decision problems. In project management, a *decision tree* may be useful for evaluating sequential project milestones. A decision problem under certainty has two elements: *action and consequence*. The decision maker's choices are the actions, while the results of those actions are the consequences. For example, in a critical path method (CPM) network planning, the choice of one task among three potential tasks in a given time slot represents a potential action. The consequences of choosing one task over another may be characterized in terms of the slack time created in the network, the cost of performing the selected task, the resulting effect on the project completion time, or the degree to which a specified performance criterion is satisfied.

If the decision is made under uncertainty, as in program evaluation review technique (PERT) network analysis, a third element, called an *event*, is introduced into the decision problem. Extending the CPM task selection example to a PERT analysis, the actions may be defined as Select Task 1, Select Task 2, and Select Task 3. The durations associated with the three possible actions can be categorized as *long task duration, medium task duration*, and *short task duration*. The actual duration of each task is uncertain. Thus, each task has some probability of exhibiting long, medium, or short durations.

Events can be identified as weather incidents: rain or no rain. The incidents of rain or no rain are uncertain. The consequences may be defined as *increased project completion time, decreased project completion time*, and *unchanged project completion time*. These consequences are also uncertain because of the probable durations of the tasks and the variable choices of the decision maker. That is, the consequences are determined partly by choice and partly by chance. The consequences also depend on which event occurs—rain or no rain.

To simplify the decision analysis, the decision elements may be summarized by using a decision table. A *decision table* shows the relationship between pairs of decision elements. Table 9.1 shows the decision table for the preceding example. In the table, each row corresponds to an event and each column corresponds to an action. The consequences appear as entries in the body of the table. The consequences have been coded as I (Increased), D (Decreased), U (Unchanged). Each event–action combination has a specific consequence associated with it.

In some decision problems, the consequences may not be unique. Thus, a consequence, which is associated with a particular event–action pair, may also be associated with another such pair. The actions included in the decision table are the only ones that the decision maker wishes to consider. Subcontracting or task elimination, for example, are other possible choices for the decision maker. The actions included in the decision problem are mutually exclusive and collectively exhaustive so that

TABLE 9.1
Decision Table for Task Selection

| | Actions | | | | | | | | |
| | Task 1 | | | Task 2 | | | Task 3 | | |
Event	Long	Medium	Short	Long	Medium	Short	Long	Medium	Short
Rain	I	I	U	I	U	D	I	I	U
No Rain	I	D	D	U	D	D	U	U	U

I = Increased project duration; D = Decreased project duration; U = Unchanged project duration.

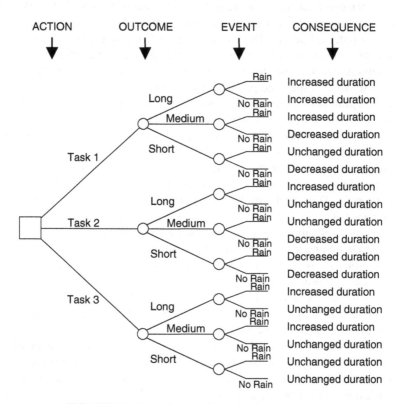

FIGURE 9.1   Decision tree for task selection example.

exactly one will be selected. The events are also mutually exclusive and collectively exhaustive.

The decision problem can also be conveniently represented as a decision tree as shown in Figure 9.1. The tree representation is particularly effective for decision problems with choices that must be made at different times over an extended period. Resource allocation decisions, for example, must be made several times during a project's life cycle. The choice of actions is shown as a fork with a separate branch for each action. The events are also represented by branches in separate fields.

To avoid confusion in very complex decision trees, the nodes for action forks are represented by squares, while the nodes for event forks are represented by circles. The basic convention for constructing a tree diagram is that the flow should be chronological from left to right. The actions are shown on the initial fork because the decision must be made before the actual event is known. The events are shown as branches in the third-stage forks. The consequence resulting from an event–action combination is shown as the end point of the corresponding path from the root of the tree.

Figure 9.1 shows six paths leading to an increase in the project duration, five paths leading to a decrease in project duration, and seven paths leading to an unchanged project duration. The total number of paths is given by:

$$P = \prod_{i=1}^{N} n_i$$

where

P = total number of paths in the decision tree

N = number of decision stages in the tree

n_i = number of branches emanating from each node in stage i

Thus, for the example in Figure 9.1, the number of paths is $P = (3)(3)(2) = 18$. As mentioned previously, some of the paths, even though they are distinct, lead to identical consequences.

Probability values can be incorporated into the decision structure as shown in Figure 9.2. Note that the selection of a task at the decision node is based on choice rather than probability. In this example, we assume that the probability of having a particular task duration is independent of whether or not it rains. In some cases, the weather sensitivity of a task may influence the duration of the task. Also, the probability of rain or no rain is independent of any other element in the decision structure.

If the items in the probability tree are interdependent, then the appropriate conditional probabilities would need to be computed. This will be the case if the duration of a task is influenced by whether or not it rains. In such a case, the probability tree should be redrawn as shown in Figure 9.3, which indicates that the weather event will need to be observed first before the task duration event can be determined. For Figure 9.3, the conditional probability of each type of duration, given that it rains or it does not rain, will need to be calculated.

The respective probabilities of the three possible consequences are shown in Figure 9.2. The probability at the end of each path is computed by multiplying the individual probabilities along the path. For example, the probability of having an increased project completion time along the first path (Task 1, Long duration, and rain) is calculated as:

$$(0.65)(0.35) = 0.2275.$$

Similarly, the probability for the second path (Task 1, Long duration, and no rain) is calculated as:

$$(0.65)(0.65) = 0.4225.$$

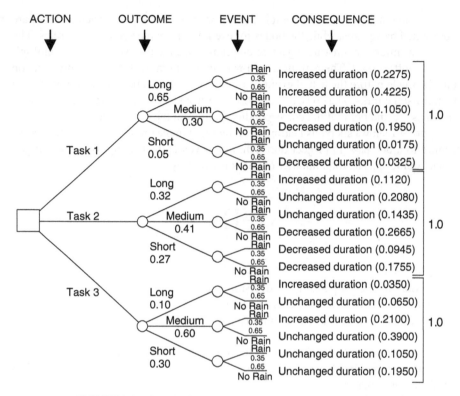

FIGURE 9.2   Probability tree diagram for task selection example.

The sum of the probabilities at the end of the paths associated with each action (choice) is equal to one as expected. Table 9.2 presents a summary of the respective probabilities of the three consequences based on the selection of each task. The probability of having an increased project duration when Task 1 is selected is calculated as:

$$\text{Probability} = 0.2275 + 0.4225 + 0.105 = 0.755$$

Likewise, the probability of having an increased project duration when Task 3 is selected is calculated as:

$$\text{Probability} = 0.035 + 0.21 = 0.245$$

If the selection of tasks at the first node is probable in nature, then the respective probabilities would be included in the calculation procedure. For example, Figure 9.4 shows a case where Task one is selected 25% of the time, Task two 45% of the time, and Task three 30% of the time. The resulting end probabilities for the three possible consequences have been revised accordingly. Note that all probabilities at the end of all the paths add up to one in this case. Table 9.3 presents the summary

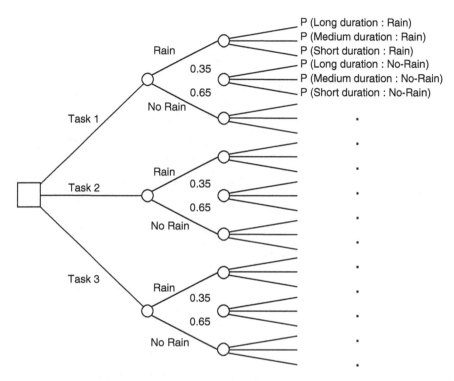

P (Long duration : Rain)
P (Medium duration : Rain)
P (Short duration : Rain)
P (Long duration : No-Rain)
P (Medium duration : No-Rain)
P (Short duration : No-Rain)

FIGURE 9.3   Probability tree for weather-dependent task durations.

TABLE 9.2
Probability Summary for Project Completion Time

Consequence	Selected task					
	Task 1		Task 2		Task 3	
Increased duration	0.2275 + 0.4225 + 0.105	0.755	0.112	0.112	0.035 + 0.21	0.245
Decreased duration	0.195 + 0.0325	0.2275	0.2665 + 0.0945 + 0.1755	0.5635	0.0	0.0
Unchanged duration	0.0175	0.0175	0.208 + 0.1435	0.3515	0.065 + 0.39 + 0.105 + 0.195	0.755
Sum of probabilities		1.0		1.0		1.0

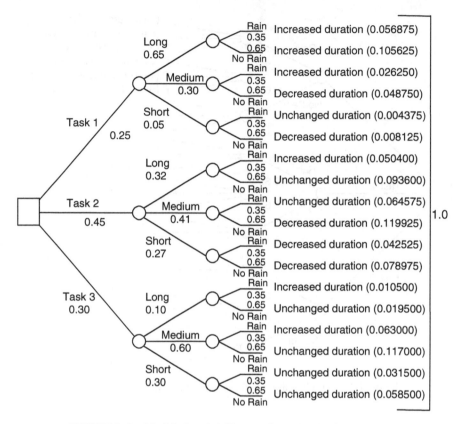

FIGURE 9.4 Modified probability tree for task selection example.

TABLE 9.3
Summary for Weather-Dependent Task Durations

Consequence	Path probabilities	Row total
Increased duration	0.056875 + 0.105625 + 0.02625 + 0.0504 + 0.0105 + 0.063	0.312650
Decreased duration	0.04875 + 0.119925 + 0.042525 + 0.078975	0.290175
Unchanged duration	0.004375 + 0.008125 + 0.0936 + 0.064575 + 0.0195 + 0.117 + 0.0315 + 0.0585	0.397175
Column total		1.0

of the probabilities of the three consequences for the case of weather-dependent task durations.

The examples presented in this chapter can be extended to other decision problems in project management that can be represented in terms of decision tables and trees. For example, resource allocation decision problems under uncertainty can be handled by appropriate decision tree models.

10 Project Economic Analysis

This chapter covers project economic analysis of industrial projects. Basic cash-flow analysis and time value of money are presented along with related topics as cost concepts, cost estimation, and cost monitoring. The time value of money is an important factor in project planning and control. This is particularly crucial for dynamic industrial projects that are subject to changes in several cost parameters due to changing technologies. Both the timing and quantity of cash flows are important for project management. The evaluation of a project alternative requires consideration of the initial investment, depreciation, taxes, inflation, economic life, salvage value, and timing of cash flows.

COST MANAGEMENT DEFINITIONS

Cost management in a project environment refers to the functions required to maintain effective financial control of the project throughout its life cycle. There are several cost concepts that influence the economic aspects of managing industrial projects. Within a given scope of analysis, there will be a combination of different types of cost factors as defined here:

Actual Cost of Work Performed: The cost actually incurred and recorded in accomplishing the work performed within a given time period.

Applied Direct Cost: The amounts recognized in the time period associated with the consumption of labor, material, and other direct resources, without regard to the date of commitment or the date of payment. These amounts are to be charged to work-in-process (WIP) when resources are actually consumed, material resources are withdrawn from inventory for use, or material resources are received and scheduled for use within 60 days.

Budgeted Cost for Work Performed: The sum of the budgets for completed work plus the appropriate portion of the budgets for level of effort and apportioned effort. Apportioned effort is effort that by itself is not readily divisible into short-span work packages but is related in direct proportion to measured effort.

Budgeted Cost for Work Scheduled: The sum of budgets for all work packages and planning packages scheduled to be accomplished (including work in process), plus the amount of level of effort and apportioned effort scheduled to be accomplished within a given period of time.

Direct Cost: Cost that is directly associated with actual operations of a project. Typical sources of direct costs are direct material costs and direct labor costs.

Direct costs are those that can be reasonably measured and allocated to a specific component of a project.

Economies of Scale: This is a term referring to the reduction of the relative weight of the fixed cost in total cost, achieved by increasing the quantity of output. Economies of scale help to reduce the final unit cost of a product and are often simply referred to as the savings due to mass production.

Estimated Cost at Completion: This refers to the sum of actual direct costs, plus indirect costs that can be allocated to a contract, plus the estimate of costs (direct and indirect) for authorized work remaining to be done.

First Cost: The total initial investment required to initiate a project or the total initial cost of the equipment needed to start the project.

Fixed Cost: Costs incurred regardless of the level of operation of a project. Fixed costs do not vary in proportion to the quantity of output. Examples of costs that make up the fixed cost of a project are administrative expenses, certain types of taxes, insurance cost, depreciation cost, and debt servicing cost. These costs usually do not vary in proportion to quantity of output.

Incremental Cost: The additional cost of changing the production output from one level to another. Incremental costs are normally variable costs.

Indirect Cost: This is a cost that is indirectly associated with project operations. Indirect costs are those that are difficult to assign to specific components of a project. An example of an indirect cost is the cost of computer hardware and software needed to manage project operations. Indirect costs are usually calculated as a percentage of a component of direct costs. For example, the direct costs in an organization may be computed as 10% of direct labor costs.

Life-Cycle Cost: This is the sum of all costs, recurring and nonrecurring, associated with a project during its entire life cycle.

Maintenance Cost: This is a cost that occurs intermittently or periodically for the purpose of keeping project equipment in good operating condition.

Marginal Cost: Marginal cost is the additional cost of increasing production output by one additional unit. The marginal cost is equal to the slope of the total cost curve or line at the current operating level.

Operating Cost: This is a recurring cost needed to keep a project in operation during its life cycle. Operating costs may consist of such items as labor, material, and energy costs.

Opportunity Cost: This refers to the cost of forgoing the opportunity to invest in a venture that, had it been pursued, would have produced an economic advantage. Opportunity costs are usually incurred due to limited resources that make it impossible to take advantage of all investment opportunities. It is often defined as the cost of the best-rejected opportunity. Opportunity costs can also be incurred due to a missed opportunity rather than due to an intentional rejection. In many cases, opportunity costs are hidden or implied because they typically relate to future events that cannot be accurately predicted.

Overhead Cost: These are costs incurred for activities performed in support of the operations of a project. The activities that generate overhead costs support the project efforts rather than contributing directly to the project goal. The handling of overhead costs varies widely from company to company. Typical overhead items are electric power cost, insurance premiums, cost of security, and inventory carrying cost.

Standard Cost: This is a cost that represents the normal or expected cost of a unit of the output of an operation. Standard costs are established in advance. They are developed as a composite of several component costs, such as direct labor cost per unit, material cost per unit, and allowable overhead charge per unit.

Sunk Cost: Sunk cost is a cost that occurred in the past and cannot be recovered under the present analysis. Sunk costs should have no bearing on the prevailing economic analysis and project decisions. Ignoring sunk costs can be a difficult task for analysts. For example, if $950,000 was spent 4 years ago to buy a piece of equipment for a technology-based project, a decision on whether or not to replace the equipment now should not consider that initial cost. But uncompromising analysts might find it difficult to ignore that much money. Similarly, an individual making a decision on selling a personal automobile would typically try to relate the asking price to what was paid for the automobile when it was acquired. This is wrong under the strict concept of sunk costs.

Total Cost: This is the sum of all the variable and fixed costs associated with a project.

Variable Cost: This cost varies in direct proportion to the level of operation or quantity of output. For example, the costs of material and labor required to make an item will be classified as variable costs as they vary with changes in level of output.

BASIC CASH-FLOW ANALYSIS

Economic analysis is performed when a choice must be made between mutually exclusive projects that compete for limited resources. The cost performance of each project will depend on the timing and levels of its expenditures. The techniques of computing cash-flow equivalence permit us to bring competing project cash flows to a common basis for comparison. The common basis depends on the prevailing interest rate. Two cash flows that are equivalent at a given interest rate will not be equivalent at a different interest rate. The basic techniques for converting cash flows from one point in time to another are presented in the following sections.

TIME VALUE OF MONEY CALCULATIONS

Cash-flow conversion involves the transfer of project funds from one point in time to another. The following notation is used for the variables involved in the conversion process:

i = interest rate per period
n = number of interest periods

P = a present sum of money
F = a future sum of money
A = a uniform end-of-period cash receipt or disbursement
G = a uniform arithmetic gradient increase in period-by-period payments or disbursements

In many cases, the interest rate used in performing economic analysis is set equal to the minimum attractive rate of return (MARR) of the decision maker. The MARR is also sometimes referred to as *hurdle rate, required internal rate of return* (IRR), *return on investment* (ROI), or *discount rate*. The value of MARR is chosen for a project based on the objective of maximizing the economic performance of the project.

CALCULATIONS WITH COMPOUND AMOUNT FACTOR

The procedure for the single payment compound amount factor finds a future amount F that is equivalent to a present amount P at a specified interest rate i after n periods. This is calculated by the following formula:

$$F = P(1 + i)^n$$

A graphic representation of the relationship between P and F is shown in Figure 10.1.

Example: A sum of $5000 is deposited in a project account and left there to earn interest for 15 years. If the interest rate per year is 12%, the compound amount after 15 years can be calculated as follows:

$$F = \$5{,}000(1 + 0.12)^{15} = \$27{,}367.85.$$

CALCULATIONS WITH PRESENT WORTH FACTOR

The present worth factor computes P when F is given. The present worth factor is obtained by solving for P in the equation for the compound amount factor. That is,

$$P = F(1 + i)^{-n}$$

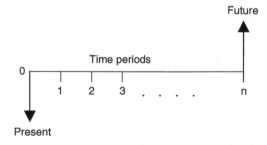

FIGURE 10.1   Single payment compound amount cash flow.

Supposing it is estimated that $15,000 would be needed to complete the implementation of a project 5 years from now, how much should be deposited in a special project fund now so that the fund would accrue to the required $15,000 exactly 5 years from now? If the special project fund pays interest at 9.2% per year, the required deposit would be:

$$P = \$15,000(1 + 0.092)^{-5} = \$9,660.03$$

CALCULATIONS WITH UNIFORM SERIES PRESENT WORTH FACTOR

The uniform series present worth factor is used to calculate the present worth equivalent, P, of a series of equal end-of-period amounts, A. Figure 10.2 shows the uniform series cash flow. The derivation of the formula uses the finite sum of the present worth values of the individual amounts in the uniform series cash flow as shown:

$$P = \sum_{t=1}^{n} A(1+i)^{-t}$$
$$= A\left[\frac{(1+i)^n - 1}{i(1+i)^n}\right]$$

Example: Suppose the sum of $12,000 must be withdrawn from an account to meet the annual operating expenses of a multiyear project. The project account pays interest at 7.5% per year compounded on an annual basis. If the project is expected to last 10 years, how much must be deposited in the project account now so that the operating expenses of $12,000 can be withdrawn at the end of every year for 10 years? The project fund is expected to be depleted to zero by the end of the last year of the project. The first withdrawal will be made 1 year after the project account is opened, and no additional deposits will be made in the account during the project life cycle. The required deposit is calculated in this way:

$$P = \$12,000\left[\frac{(1+0.075)^{10} - 1}{0.075(1+0.075)^{10}}\right]$$
$$= \$82,368.92$$

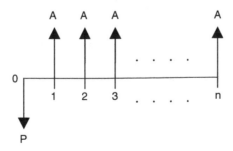

FIGURE 10.2 Uniform series cash flow.

CALCULATIONS WITH UNIFORM SERIES CAPITAL RECOVERY FACTOR

The capital recovery formula is used to calculate the uniform series of equal end-of-period payments, A, that are equivalent to a given present amount P. This is the converse of the uniform series present amount factor. The equation for the uniform series capital recovery factor is obtained by solving for A in the uniform series present amount factor. That is,

$$A = P\left[\frac{i(1+i)^n}{(1+i)^n - 1}\right]$$

Example: Suppose a piece of equipment needed to launch a project must be purchased at a cost of \$50,000. The entire cost is to be financed at 13.5% per year and repaid on a monthly installment schedule over 4 years. It is desired to calculate what the monthly loan payments will be. It is assumed that the first loan payment will be made exactly 1 month after the equipment is financed. If the interest rate of 13.5% per year is compounded monthly, then the interest rate per month will be 13.5/12 = 1.125% per month. The number of interest periods over which the loan will be repaid is 4(12) = 48 months. Consequently, the monthly loan payments are calculated to be:

$$A = \$50,000\left[\frac{0.01125(1+0.01123)^{48}}{(1+0.01125)^{48} - 1}\right]$$

$$= \$1353.82$$

CALCULATIONS WITH UNIFORM SERIES COMPOUND AMOUNT FACTOR

The series compound amount factor is used to calculate a single future amount that is equivalent to a uniform series of equal end-of-period payments. The cash flow is shown in Figure 10.3. Note that the future amount occurs at the same point in time as the last amount in the uniform series of payments. The factor is derived as shown:

$$F = \sum_{t=1}^{n} A(1+i)^{n-t}$$

$$= A\left[\frac{(1+i)^n - 1}{i}\right]$$

Example: If equal end-of-year deposits of \$5000 are made to a project fund paying 8% per year for 10 years, how much can be expected to be available for withdrawal from the account for capital expenditure immediately after the last deposit is made?

$$F = \$5000\left[\frac{(1+0.08)^{10} - 1}{0.08}\right]$$

$$= \$72,432.50$$

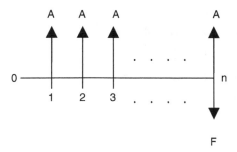

FIGURE 10.3   Uniform series compound amount cash flow.

CALCULATIONS WITH UNIFORM SERIES SINKING FUND FACTOR

The sinking fund factor is used to calculate the uniform series of equal end-of-period amounts, A, that are equivalent to a single future amount F. This is the reverse of the uniform series compound amount factor. The formula for the sinking fund is obtained by solving for A in the formula for the uniform series compound amount factor. That is,

$$A = F\left[\frac{i}{(1+i)^n -1}\right]$$

Example: How large are the end-of-year equal amounts that must be deposited into a project account so that a balance of $75,000 will be available for withdrawal immediately after the twelfth annual deposit is made? The initial balance in the account is zero at the beginning of the first year. The account pays 10% interest per year. Using the formula for the sinking fund factor, the required annual deposits are:

$$A = \$75,000\left[\frac{0.10}{(1+0.10)^{12} -1}\right]$$
$$= \$3,507.25$$

CALCULATIONS WITH CAPITALIZED COST FORMULA

Capitalized cost refers to the present value of a single amount that is equivalent to a perpetual series of equal end-of-period payments. This is an extension of the series present worth factor with an infinitely large number of periods. This is shown graphically in Figure 10.4.

Using the limit theorem from calculus as n approaches infinity, the series present worth factor reduces to the following formula for the capitalized cost:

$$P = \frac{A}{i}$$

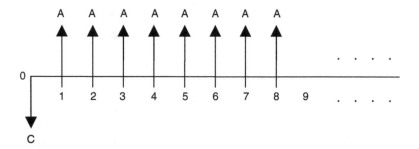

FIGURE 10.4 Capitalized cost cash flow.

Example: How much should be deposited in a general fund to service a recurring public service project to the tune of $6500 per year forever if the fund yields an annual interest rate of 11%? Using the capitalized cost formula, the required one-time deposit to the general fund is:

$$P = \frac{\$6500}{0.11}$$
$$= \$59,090.91$$

ARITHMETIC GRADIENT SERIES

The gradient series cash flow involves an increase of a fixed amount in the cash flow at the end of each period. Thus, the amount at a given point in time is greater than the amount at the preceding period by a constant amount. This constant amount is denoted by G. Figure 10.5 shows the basic gradient series in which the base amount at the end of the first period is zero. The size of the cash flow in the gradient series at the end of period t is calculated as

$$A_t = (t-1)G, \quad t = 1, 2, ..., n$$

The total present value of the gradient series is calculated by using the present amount factor to convert each individual amount from time t to time 0 at an interest rate of $i\%$ per period and then summing up the resulting present values. The finite summation reduces to a closed form as shown:

$$P = \sum_{t=1}^{n} A_t (1+i)^{-t}$$
$$= G\left[\frac{(1+i)^n - (1+ni)}{i^2(1+i)^n}\right]$$

Example: The cost of supplies for a 10-year project increases by $1500 every year starting at the end of year two. There is no cost for supplies at the end of the first year. If interest rate is 8% per year, determine the present amount that must be set aside at time zero to take care of all the future supplies expenditures. We have $G = 1500$, $i = 0.08$, and $n = 10$. Using the arithmetic gradient formula, we obtain

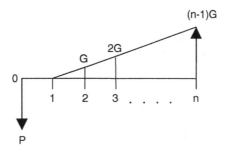

FIGURE 10.5   Arithmetic gradient cash flow with zero base amount.

$$P = 1500\left[\frac{1-\left(1+10(0.08)\right)\left(1+0.08\right)^{-10}}{(0.08)^2}\right]$$
$$= \$1500(25.9768)$$
$$= \$38,965.20$$

In many cases, an arithmetic gradient starts with some base amount at the end of the first period and then increases by a constant amount thereafter. The nonzero base amount is denoted as A_1. Figure 10.6 shows this type of cash flow.

The calculation of the present amount for such cash flows requires breaking the cash flow into a uniform series cash flow of amount A_1 and an arithmetic gradient cash flow with zero base amount. The uniform series present worth formula is used to calculate the present worth of the uniform series portion, while the basic gradient series formula is used to calculate the gradient portion. The overall present worth is then calculated as follows:

$$P = P_{\text{uniform series}} + P_{\text{gradient series}}$$
$$= A_1\left[\frac{(1+i)^n - 1}{i(1+i)^n}\right] + G\left[\frac{(1+i)^n - (1+ni)}{i^2(1+i)^n}\right]$$

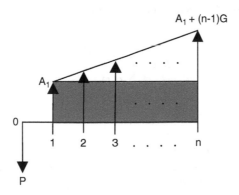

FIGURE 10.6   Arithmetic gradient cash flow with nonzero base amount.

INTERNAL RATE OF RETURN

The internal rate of return (IRR) for a cash flow is defined as the interest rate that equates the future worth at time n or present worth at time 0 of the cash flow to zero. If we let $i*$ denote the internal rate of return, then we have:

$$FW_{t=n} = \sum_{t=0}^{n} (\pm A_t)(1+i*)^{n-t} = 0$$

$$PW_{t=0} = \sum_{t=0}^{n} (\pm A_t)(1+i*)^{-t} = 0$$

where $+$ is used in the summation for positive cash-flow amounts or receipts and $-$ is used for negative cash-flow amounts or disbursements. A_t denotes the cash-flow amount at time t, which may be a receipt $(+)$ or a disbursement $(-)$. The value of $i*$ is referred to as *discounted cash flow rate of return, internal rate of return,* or *true rate of return*. The aforementioned procedure essentially calculates the net future worth or the net present worth of the cash flow. That is:

Net Future Worth = Future Worth of Receipts – Future Worth of Disbursements

that is,

$$NFW = FW_{receipts} - FW_{disbursements}$$

Net present worth = Present worth of receipts – Present worth of disbursements

that is,

$$NPW = PW_{receipts} - PW_{disbursements}$$

Setting the NPW or NFW equal to zero and solving for the unknown variable i determines the internal rate of return of the cash flow.

BENEFIT–COST RATIO ANALYSIS

The benefit–cost ratio of a cash flow is the ratio of the present worth of benefits to the present worth of costs. This is defined as follows:

$$B/C = \frac{\sum_{t=0}^{n} B_t (1+i)^{-t}}{\sum_{t=0}^{n} C_t (1+i)^{-t}}$$

$$= \frac{PW_{benefits}}{PW_{costs}}$$

where B_t is the benefit (receipt) at time t and C_t is the cost (disbursement) at time t. If the benefit–cost ratio is greater than one, then the investment is acceptable. If the ratio is less than one, the investment is not acceptable. A ratio of one indicates a break-even situation for the project.

SIMPLE PAYBACK PERIOD

Payback period refers to the length of time it will take to recover an initial investment. The approach does not consider the impact of the time value of money. Consequently, it is not an accurate method of evaluating the worth of an investment. However, it is a simple technique that is used widely to perform a "quick-and-dirty" assessment of investment performance. Another limitation of the technique is that it considers only the initial cost. Other costs that may occur after time zero are not included in the calculation. The payback period is defined as the smallest value of n (n_{min}) that satisfies the following expression:

$$\sum_{t=1}^{n_{min}} R_t \geq C$$

where R_t is the revenue at time t and C_0 is the initial investment. The procedure calls for a simple addition of the revenues period by period until enough total has been accumulated to offset the initial investment.

Example: An organization is considering installing a new computer system that will generate significant savings in material and labor requirements for order processing. The system has an initial cost of $50,000. It is expected to save the organization $20,000 a year. The system has an anticipated useful life of 5 years with a salvage value of $5000. Determine how long it would take for the system to pay for itself from the savings it is expected to generate. As the annual savings are uniform, we can calculate the payback period by simply dividing the initial cost by the annual savings. That is,

$$n_{min} = \frac{\$50,000}{\$20,000}$$
$$= 2.5 \text{ years.}$$

Note that the salvage value of $5000 is not included in the above calculation because the amount is not realized until the end of the useful life of the asset (i.e., after 5 years). In some cases, it may be desired to consider the salvage value. In that case, the amount to be offset by the annual savings will be the net cost of the asset. In that case, we would have the following:

$$n_{min} = \frac{\$50,000 - \$5000}{\$20,000}$$
$$= 2.25 \text{ years.}$$

If there are tax liabilities associated with the annual savings, those liabilities must be deducted from the savings before the payback period is calculated.

DISCOUNTED PAYBACK PERIOD

In this book, we introduce the *discounted payback period* approach, in which the revenues are reinvested at a certain interest rate. The payback period is determined when enough money has been accumulated at the given interest rate to offset the initial cost as well as other interim costs. In this case, the calculation is done by the following expression:

$$\sum_{t=1}^{n_{min}} R_t \left(1+i\right)^{n_{min}-1} \geq \sum_{t=0}^{n_{min}} C_t \,.$$

Example: A new solar cell unit is to be installed in an office complex at an initial cost of $150,000. It is expected that the system will generate annual cost savings of $22,500 on the electricity bill. The solar cell unit will need to be overhauled every 5 years at a cost of $5,000 per overhaul. If the annual interest rate is 10%, find the discounted payback period for the solar cell unit considering the time value of money. The costs of overhaul are to be considered in calculating the discounted payback period.

Solution: Using the single payment compound amount factor for one period iteratively, the following solution is obtained as follows:

Time	Cumulative savings
1	$22,500
2	$22,500 + $22,500 (1.10)^1 = $47,250
3	$22,500 + $47,250 (1.10)^1 = $74,475
4	$22,500 + $74,475 (1.10)^1 = $104,422.50
5	$22,500 + $104,422.50 (1.10)^1 − $5,000 = $132,364.75
6	$22,500 + $132,364.75 (1.10)^1 = $168,101.23

The initial investment is $150,000. By the end of period 6, we will have accumulated $168,101.23, more than the initial cost. Interpolating between period 5 and period 6 results in n_{min} of 5.49 years. That is, it will take 5.5 years to recover the initial investment. The calculation is shown as follows:

$$n_{min} = 5 + \frac{150,000 - 132,364.75}{168,101.25 - 132,364.75}(6-5)$$
$$= 5.49$$

TIME REQUIRED TO DOUBLE INVESTMENT

It is sometimes of interest to determine how long it will take a given investment to reach a certain multiple of its initial level. The "Rule of 72" is one simple approach to calculating the time required to for an investment to double in value, at a given interest rate per period. The Rule of 72 gives the following formula for estimating the time required:

$$n = \frac{72}{i}$$

where i is the interest rate expressed in percentage. Referring to the single payment compound amount factor, we can set the future amount equal to twice the present amount and then solve for n. That is, $F = 2P$. Thus,

$$2P = P(1+i)^n.$$

Solving for n in this equation yields an expression for calculating the exact number of periods required to double P:

$$n = \frac{\ln(2)}{\ln(1+i)},$$

where i is the interest rate expressed in decimals. In general, the length of time it would take to accumulate m multiples of P is expressed as:

$$n = \frac{\ln(m)}{\ln(1+i)},$$

where m is the desired multiple. For example, at an interest rate of 5% per year, the time it would take an amount P to double in value ($m = 2$) is 14.21 years. This, of course, assumes that the interest rate will remain constant throughout the planning horizon. Table 10.1 presents a tabulation of the values calculated from both approaches. Figure 10.7 shows a graphical comparison of the Rule of 72 to the exact calculation.

EFFECTS OF INFLATION ON INDUSTRIAL PROJECT COSTING

Inflation can be defined as the decline in purchasing power of money, and as such, is a major player in the financial and economic analysis of projects. Multiyear projects are particularly subject to the effects of inflation. Some of the most common causes of inflation include the following:

- An increase in the amount of currency in circulation
- A shortage of consumer goods
- An escalation of the cost of production
- An arbitrary increase in prices set by resellers

The general effects of inflation are felt in terms of an increase in the prices of goods and a decrease in the worth of currency. In cash-flow analysis, return on investment (ROI) for a project will be affected by time value of money as well as inflation. The real interest rate (d) is defined as the desired rate of return in the absence of inflation. When we talk of "today's dollars" or "constant dollars," we are referring to the use of the real interest rate. The combined interest rate (i) is the rate of return combining the real interest rate and the inflation rate. If we denote the inflation rate as j, then the relationship between the different rates can be expressed as:

$$1 + i = (1 + d)(1 + j)$$

Thus, the combined interest rate can be expressed as follows:

$$i = d + j + dj$$

TABLE 10.1
Evaluation of the Rule of 72

i (%)	n (Rule of 72)	n (Exact value)
0.25	288.00	277.61
0.50	144.00	138.98
1.00	72.00	69.66
2.00	36.00	35.00
5.00	14.20	17.67
8.00	9.00	9.01
10.00	7.20	7.27
12.00	6.00	6.12
15.00	4.80	4.96
18.00	4.00	4.19
20.00	3.60	3.80
25.00	2.88	3.12
30.00	2.40	2.64

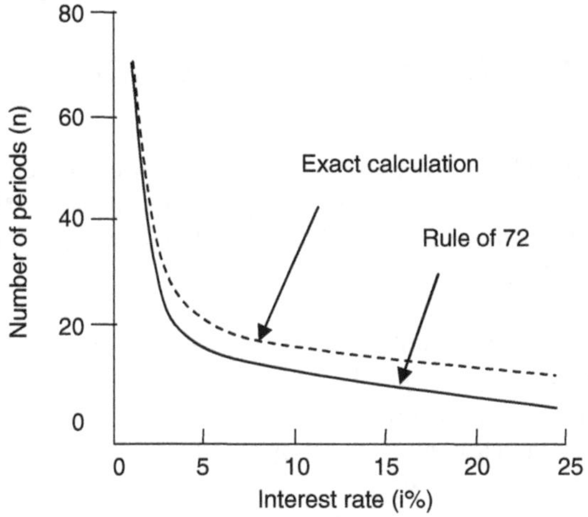

FIGURE 10.7   Evaluation of investment life for double return.

Note that if $j = 0$ (i.e., no inflation), then $i = d$. We can also define commodity escalation rate (g) as the rate at which individual commodity prices escalate. This may be greater than or less than the overall inflation rate. In practice, several measures are used to convey inflationary effects. Some of these are the Consumer Price Index, the Producer Price Index, and the Wholesale Price Index. A "market basket" rate is defined as the estimate of inflation based on a weighted average of the annual rates of change in the costs of a wide range of representative commodities. A "then-current" cash flow is a cash flow that explicitly incorporates the impact of inflation. A "constant worth" cash flow is a cash flow that does not incorporate the effect of

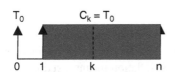
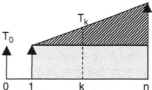

FIGURE 10.8 Cash flows for effects of inflation.

inflation. The real interest rate, d, is used for analyzing constant worth cash flows. Figure 10.8 shows constant worth and then-current cash flows.

The then-current cash flow in the figure is the equivalent cash flow considering the effect of inflation. C_k is what it would take to buy a certain "basket" of goods after k time periods if there was no inflation. T_k is what it would take to buy the same "basket" in k time period if inflation were taken into account. For the constant worth cash flow, we have:

$$C_k = T_0, k = 1, 2, ..., n$$

And for the then-current cash flow, we have:

$$T_k = T_0(1 + j)^k, k = 1, 2, ..., n$$

where j is the inflation rate. If $C_k = T_0 = \$100$ under the constant worth cash flow, then we have \$100 worth of buying power. If we are using the commodity escalation rate, g, then we will have:

$$T_k = T_0(1 + g)^k, k = 1, 2, ..., n.$$

Thus, a then-current cash flow may increase based on both a regular inflation rate (j) and a commodity escalation rate (g). We can convert a then-current cash flow to a constant worth cash flow by using the following relationship:

$$C_k = T_k(1 + j)^{-k}, k = 1, 2, ..., n.$$

If we substitute T_k from the commodity escalation cash flow into the expression for C_k, we get the following:

$$\begin{aligned}
C_k &= T_k (1+j)^{-k} \\
&= T (1+g)^k (1+j)^{-k} \\
&= T_0 \left[(1+g)/(1+j) \right]^k, \quad k = 1, 2, \cdots, n
\end{aligned}$$

Note that if $g = 0$ and $j = 0$, the $C_k = T_0$. That is, there is no inflationary effect. We can now define the effective commodity escalation rate (v):

$$v = [(1 + g)/(1 + j)] - 1$$

The commodity escalation rate (g) can be expressed as follows:

$$g = v + j + vj.$$

Inflation can have a significant impact on the financial and economic aspects of an industrial project. Inflation may be defined, in economic terms, as the increase in the amount of currency in circulation. To a producer, inflation means a sudden increase in the cost of items that serve as inputs for the production process (equipment, labor, materials, etc.). To the retailer, inflation implies an imposed higher cost of finished products. To an ordinary citizen, inflation portends a noticeable escalation of prices of consumer goods. All these aspects are intertwined in a project management environment.

The amount of money supply, as a measure of a country's wealth, is controlled by the government. When circumstances dictate such action, governments often feel compelled to create more money or credit to take care of old debts and pay for social programs. When money is generated at a faster rate than the growth of goods and services, it becomes a surplus commodity and its value (i.e., purchasing power) will fall. This means that there will be too much money available to buy only a few goods and services. When the purchasing power of a currency falls, each individual in a product's life cycle (that is, each person or entity that spends money on a product throughout its life cycle, from production through disposal) has to use more of the currency in order to obtain the product. Some of the classic concepts of inflation are discussed:

1. In *cost-driven* or *cost-push inflation*, increases in producer's costs are passed on to consumers. At each stage of the product's journey from producer to consumer, prices are escalated disproportionately in order to make a good profit. The overall increase, in the product's price is directly proportional to the number of intermediaries it encounters on its way to the consumer.
2. In *demand-driven* or *demand-pull inflation*, *excessive* spending power of consumers forces an upward trend in prices. This high spending power is usually achieved at the expense of savings. The law of supply and demand dictates that the more the demand, the higher the price. This results in *demand-driven* or *demand-pull inflation*.
3. Impact of international economic forces can induce inflation on a local economy. Trade imbalances and fluctuations in currency values are notable examples of international inflationary factors.
4. In *wage-driven* or *wage-push inflation*, the increasing base wages of workers generate more disposable income and, hence, higher demands for goods and services. The high demand, consequently, creates a pull on prices. Coupled with this, employers pass the additional wage cost on to consumers through higher prices. This type of inflation is very difficult to contain because wages set by union contracts and prices set by producers almost never fall.
5. Easy availability of credit leads consumers to "buy now and pay later," thereby creating another opportunity for inflation. This is a dangerous type of inflation because the credit not only pushes prices up but it also leaves consumers with less money later to pay for the credit. Eventually, many

credits become uncollectible debts, which may then drive the economy toward recession.

6. Deficit spending results in an increase in money supply and, thereby, creates less room for each dollar to get around. The popular saying that indicates that "a dollar does not go far anymore" simply refers to inflation in laymen's terms. The different levels of inflation may be categorized as discussed in the following sections.

MILD INFLATION

When inflation is mild (at 2%–4%) the economy actually prospers. Producers strive to produce at full capacity in order to take advantage of the high prices to the consumer. Private investments tend to be brisk, and more jobs become available. However, the good fortune may only be temporary. Prompted by the prevailing success, employers are tempted to seek larger profits and workers begin to ask for higher wages. They cite their employer's prosperous business as a reason to bargain for bigger shares of the business profit. So, we end up with a vicious cycle in which the producer asks for higher prices, the unions ask for higher wages, and inflation starts an upward trend.

MODERATE INFLATION

Moderate inflation occurs when prices increase at 5%–9%. Consumers start purchasing more as a hedge against inflation. They would rather spend their money now than watch it decline further in purchasing power. The increased market activity serves to fuel further inflation.

SEVERE INFLATION

Severe inflation is indicated by price escalations of 10% or more. Double-digit inflation implies that prices rise much faster than wages do. Debtors tend to be the ones who benefit from this level of inflation because they repay debts with money that is less valuable than when they borrowed.

HYPERINFLATION

When each price increase signals an increase in wages and costs, which again sends prices further up, the economy has reached a stage of malignant galloping inflation or hyperinflation. Rapid and uncontrollable inflation destroys the economy. The currency becomes economically useless as the government prints it excessively to pay for obligations.

Inflation can affect any industrial project in terms of raw materials procurement, salaries and wages, and/or cost tracking dilemmas. Some effects are immediate and easily observable while others are subtle and pervasive. Whatever form it takes, inflation must be taken into account in long-term project planning and control. Large projects, especially, may be adversely affected by the effects of inflation in terms of cost overruns and poor resource utilization. Managers should note that the level of inflation will determine the severity of the impact on projects.

BREAK-EVEN ANALYSIS

Break-even analysis refers to the determination of the balanced performance level where project income is equal to project expenditure. The total cost of an operation is expressed as the sum of the fixed and variable costs with respect to output quantity. That is,

$$TC(x) = FC + VC(x)$$

where x is the number of units produced, $TC(x)$ is the total cost of producing x units, FC is the total fixed cost, and $VC(x)$ is the total variable cost associated with producing x units. The total revenue resulting from the sale of x units is defined as

$$TR(x) = px$$

where p is the price per unit. The profit due to the production and sale of x units of the product is calculated as

$$P(x) = TR(x) - TC(x).$$

The break-even point of an operation is defined as the value of a given parameter that will result in neither profit nor loss. The parameter of interest may be the number of units produced, the number of hours of operation, the number of units of a resource type allocated, or any other measure of interest. At the break-even point, we have the following relationship:

$$TR(x) = TC(x) \text{ or } P(x) = 0.$$

In some cases, there may be a known mathematical relationship between cost and the parameter of interest. For example, there may be a linear cost relationship between the total cost of a project and the number of units produced. The cost expressions facilitate a straightforward break-even analysis. Figure 10.9 shows an example of a break-even point for a single project. Figure 10.10 shows examples of multiple break-even points that exist when multiple projects are compared. When two project alternatives are compared, the break-even point refers to the point of indifference between the two alternatives. In Figure 10.10, x_1 represents the point where projects A and B are equally desirable, x_2 represents where A and C are equally desirable, and x_3 represents where B and C are equally desirable. The figure shows that, if we are operating below a production level of x_2 units, then project C is the preferred project among the three. If we are operating at a level more than x_2 units, then project A is the best choice.

Example: Three project alternatives are being considered for producing a new product. The required analysis involves determining which alternative should be selected on the basis of how many units of the product are produced per year. Based on past records, there is a known relationship between the number of units produced per year, x, and the net annual profit, $P(x)$, from each alternative. The level of production

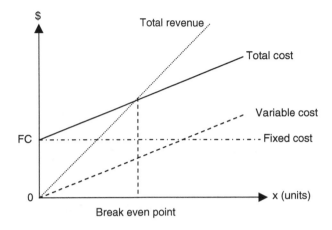

FIGURE 10.9 Break-even point for a single project.

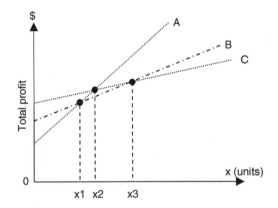

FIGURE 10.10 Break-even points for multiple projects.

is expected to be between 0 and 250 units per year. The net annual profits (in thousands of dollars) are given for each alternative:

Project A: $P(x) = 3x - 200$
Project B: $P(x) = x$
Project C: $P(x) = (1/50)x^2 - 300$.

This problem can be solved mathematically by finding the intersection points of the profit functions and evaluating the respective profits over the given range of product units. It can also be solved by a graphical approach. Figure 10.11 shows a plot of the profit functions. Such a plot is called a break-even chart. The plot shows that Project B should be selected if between 0 and 100 units are to be produced, Project A should be selected if between 100 and 178.1 units (178 physical units) are to be produced, and Project C should be selected if more than 178 units are to be produced.

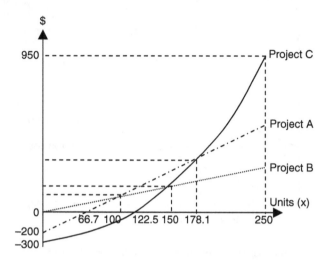

FIGURE 10.11 Plot of profit functions.

It should be noted that, if less than 66.7 units (66 physical units) are produced, Project A will generate a net loss rather than a net profit. Similarly, Project C will generate losses if less than 122.5 units (122 physical units) are produced.

PROFIT RATIO ANALYSIS

Break-even charts offer opportunities for several different types of analysis. In addition to the break-even points, other measures of worth or criterion measures may be derived from the charts. A measure called the *profit ratio* (Badiru, 1996), is presented here for the purpose of obtaining a further comparative basis for competing projects. A profit ratio is defined as the ratio of the profit area to the sum of the profit and loss areas in a break-even chart. That is,

$$\text{Profit ratio} = \frac{\text{Area of profit region}}{\text{Area of profit region} + \text{Area of loss region}}$$

For example, suppose that the expected revenue and the expected total cost associated with a project are given, respectively, by the following expressions:

$$R(x) = 100 + 10x$$
$$TC(x) = 2.5x + 250$$

where x is the number of units produced and sold from the project. Figure 10.12 shows the break-even chart for the project. The break-even point is shown to be 20 units. Net profits are realized from the project if more than 20 units a213 re produced, and net losses are realized if less than 20 units are produced. It should be noted that the revenue function in Figure 10.12 represents an unusual case, in which a revenue of $100 is realized when zero units are produced.

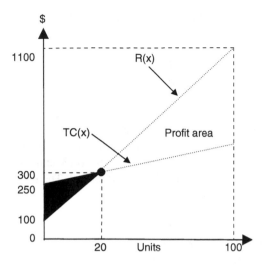

FIGURE 10.12   Area of profit versus area of loss.

Suppose it is desired to calculate the profit ratio for this project if the number of units that can be produced is limited to between 0 and 100 units. From Figure 10.12, the surface area of the profit region and the area of the loss region can be calculated by using the standard formula for finding the area of a triangle: Area = (½)(Base)(Height). Using this formula, we have the following:

$$\text{Area of profit region} = \frac{1}{2}(\text{Base})(\text{Height})$$

$$= \frac{1}{2}(1100 - 500)(100 - 20)$$

$$= 24,000 \text{ square units}$$

$$\text{Area of loss region} = \frac{1}{2}(\text{Base})(\text{Height})$$

$$= \frac{1}{2}(250 - 100)(20)$$

$$= 1500 \text{ square units}$$

Thus, the profit ratio is computed as follows:

$$\text{Profit Ratio} = 24,000/(24,000 + 1,500) = 0.9411 \equiv 94.11\%$$

The profit ratio may be used as a criterion for selecting among project alternatives. If this is done, the profit ratios for all the alternatives must be calculated over the same values of the independent variable. The project with the highest profit ratio will be selected as the desired project. For example, Figure 10.13 presents the break-even chart for an alternate project, say Project II. It can be seen that both the revenue and cost functions for the project are nonlinear. The revenue and cost are defined as follows:

$$R(x) = 160x - x2$$
$$TC(x) = 500 + x2.$$

If the cost and/or revenue functions for a project are not linear, the areas bounded by the functions may not be easily determined. For those cases, it may be necessary to use techniques such as definite integrals to find the areas. Figure 10.13 indicates that the project generates a loss if less than 3.3 units (3 actual units) are produced or if more than 76.8 units (76 actual units) are produced. The respective profit and loss areas on the chart are calculated as:

Area 1 (Loss) = 802.80 unit-dollars
Area 2 (Profit) = 132,272.08 unit-dollars
Area 3 (Loss) = 48,135.98 unit-dollars

Consequently, the profit ratio for Project II is computed as

$$\text{Profit ratio} = \frac{\text{Total area of profit region}}{\text{Total area of profit region} + \text{Total area of loss region}}$$

$$= \frac{132,272.08}{802.76 + 132,272.08 + 48,135.98}$$

$$= 72.99\%$$

The profit ratio approach evaluates the performance of each alternative over a specified range of operating levels. Most of the existing evaluation methods use single-point analysis with the assumption that the operating condition is fixed at a

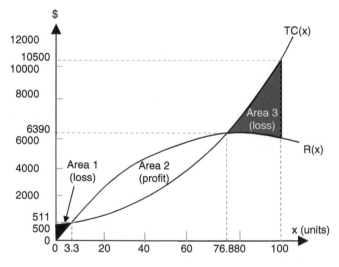

FIGURE 10.13   Break-even chart for revenue and cost functions.

given production level. The profit ratio measure allows an analyst to evaluate the net yield of an alternative, given that the production level may shift from one level to another. An alternative, for example, may operate at a loss for most of its early life, but it may generate large incomes to offset those losses in its later stages. Conventional methods cannot easily capture this type of transition from one performance level to another. In addition to being used to compare alternate projects, the profit ratio may also be used for evaluating the economic feasibility of a single project. In such a case, a decision rule may be developed, such as the following:

If profit ratio is greater than 75%, accept the project.
If profit ratio is less than or equal to 75%, reject the project.

INDUSTRIAL PROJECT COST ESTIMATION

Cost estimation and budgeting help establish a strategy for allocating resources in project planning and control. Based on the desired level of accuracy, there are three major categories of cost estimation for budgeting: *order-of-magnitude estimates*, *preliminary cost estimates*, and *detailed cost estimates*. Order-of-magnitude cost estimates are usually gross estimates based on the experience and judgment of the estimator. They are sometimes called "ballpark" figures. These estimates are typically made without a formal evaluation of the details involved in the project. The level of accuracy associated with order-of-magnitude estimates can range from −50% to +50% of the actual cost. These estimates provide a quick way of getting cost information during the initial stages of a project. The estimation range is summarized as follows:

50% (Actual cost) ≤ Order-of-magnitude estimate ≤ 150% (Actual cost).

Preliminary cost estimates are also gross estimates, but with a higher level of accuracy. In developing preliminary cost estimates, more attention is paid to some selected details of the project. An example of a preliminary cost estimate is the estimation of expected labor cost. Preliminary estimates are useful for evaluating project alternatives before final commitments are made. The level of accuracy associated with preliminary estimates can range from −20% to +20% of the actual cost, as shown:

80% (Actual cost) ≤ Preliminary estimate ≤ 120% (Actual cost).

Detailed cost estimates are developed after careful consideration is given to all the major details of a project. Considerable time is typically needed to obtain detailed cost estimates. Because of the amount of time and effort needed to develop detailed cost estimates, the estimates are usually developed after a firm commitment has been made that the project will take off. Detailed cost estimates are important for evaluating actual cost performance during the project. The level of accuracy associated with detailed estimates normally ranges from −5% to +5% of the actual cost.

95% (Actual cost) ≤ Detailed cost ≤ 105% (Actual cost)

There are two basic approaches to generating cost estimates. The first one is a variant approach, in which cost estimates are based on variations of previous cost records. The other approach is the generative cost estimation, in which cost estimates are developed from scratch without taking previous cost records into consideration.

OPTIMISTIC AND PESSIMISTIC COST ESTIMATES

Using an adaptation of the PERT formula, we can combine optimistic and pessimistic cost estimates. If O = optimistic cost estimate, M = most likely cost estimate, and P = pessimistic cost estimate, the estimated cost can be stated as follows:

$$E[C] = \frac{O + 4M + P}{6}$$

and the cost variance can be estimated as follows:

$$V[C] = \left[\frac{P - O}{6}\right]^2.$$

PROJECT BUDGET ALLOCATION

Project budget allocation involves sharing limited resources among competing tasks in a project. The budget allocation process serves the following purposes:

- A plan for resource expenditure
- A project selection criterion
- A projection of project policy
- A basis for project control
- A performance measure
- A standardization of resource allocation
- An incentive for improvement

TOP-DOWN BUDGETING

Top-down budgeting involves collecting data from upper-level sources such as top and middle managers. The figures supplied by the managers may come from their personal judgment, past experience, or past data on similar project activities. The cost estimates are passed to lower-level managers, who then break the estimates down into specific work components within the project. These estimates may, in turn, be given to line managers, supervisors, and lead workers to continue the process until individual activity costs are obtained. Thus, top management provides the global budget, while the functional level worker provides specific budget requirements for project items.

BOTTOM-UP BUDGETING

In this method, elemental activities, and their schedules, descriptions, and labor skill requirements are used to construct detailed budget requests. Line workers familiar with specific activities are asked to provide cost estimates and then make estimates

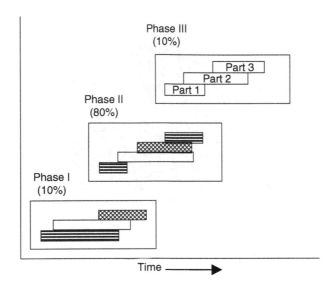

FIGURE 10.14 Budgeting by project phases.

for each activity in terms of labor time, materials, and machine time. The estimates are then converted to an appropriate cost basis. The dollar estimates are combined into composite budgets at each successive level up the budgeting hierarchy. If estimate discrepancies develop, they can be resolved through the intervention of senior management, middle management, functional managers, project manager, accountants, or standard cost consultants. Figure 10.14 shows the breaking down of a project into phases and parts in order to facilitate bottom-up budgeting and improve both schedule and cost control.

Elemental budgets may be developed on the basis of the timed progress of each part of the project. When all the individual estimates are gathered, we can obtain a composite budget estimate. Figure 10.15 and 10.16 show an example of the various components that may be involved in an overall budget. The bar chart appended to a segment of the pie chart indicates the individual cost components making up that particular segment. To further aid in the process, analytical tools such as learning curve analysis, work sampling, and statistical estimation may be employed in the cost estimation and budgeting processes.

COST MONITORING

As a project progresses, costs can be monitored and evaluated to identify areas of unacceptable cost performance. Figure 10.17 shows a plot of cost versus time for projected cost and actual cost. The plot permits a quick identification of the points at which cost overruns occur in a project.

Plots similar to those presented earlier may be used to evaluate cost, schedule, and time performance of a project. An approach similar to the profit ratio presented earlier may be used along with the plot to evaluate the overall cost performance of

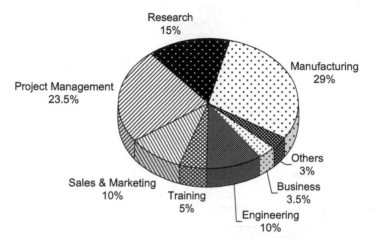

FIGURE 10.15   Pie chart of budget distribution.

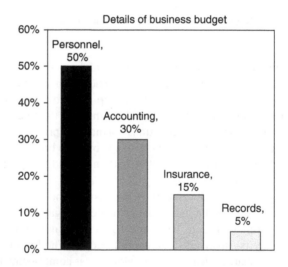

FIGURE 10.16   Bar chart of budget and distribution.

a project over a specified planning horizon. Presented below is a formula for cost performance index (CPI):

$$CPI = \frac{\text{Area of cost benefit}}{\text{Area of cost benefit} + \text{area of cost overrun}}.$$

As in the case of the profit ratio, CPI may be used to evaluate the relative performances of several project alternatives or to evaluate the feasibility and acceptability of an individual alternative. In Figure 10.18, we present another cost-monitoring tool, referred to as a cost-control pie chart. The chart is used to track the percentage of the cost going into a specific component of a project. Control limits can be included

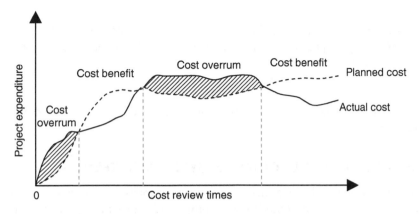

FIGURE 10.17   Evaluation of actual and projected cost.

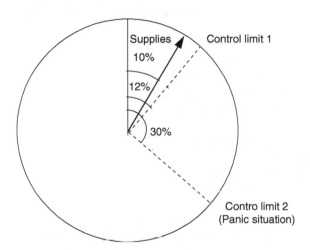

FIGURE 10.18   Cost-control pie chart.

in the pie chart to identify costs that have become out of control. The example in Figure 10.18 shows that 10% of total cost is tied up in supplies. The control limit is located at 12% of total cost. Hence, the supplies expenditure is within control (so far, at least).

PROJECT BALANCE TECHNIQUE

One other approach to monitoring cost performance is the project balance technique. The technique helps in assessing the economic state of a project at a desired point in time in the life cycle of the project. It calculates the net cash flow of a project up to a given point in time. The project balance is calculated as follows:

$$B(i)_t = S_t - P(1+i)^t + \sum_{k=1}^{t} PW_{\text{income}}(i)_k,$$

where

$B(i)_t$ = project balance at time t at an interest rate of $i\%$ per period
$PW_{income}(i)_k$ = present worth of net income from the project up to time k
P = initial cost of the project
S_t = salvage value at time t

The project balance at time t gives the net loss or net profit associated with the project up to that time.

COST AND SCHEDULE CONTROL SYSTEMS CRITERIA

Contract management involves the process by which goods and services are acquired, utilized, monitored, and controlled in a project. Contract management addresses the contractual relationships from the initiation of a project to the completion of the project (i.e., completion of services and/or hand over of deliverables). Some of the important aspects of contract management with which it is essential to be familiar include

- Principles of contract law
- Bidding process and evaluation
- Contract and procurement strategies
- Selection of source and contractors
- Negotiation
- Worker safety considerations
- Product liability
- Uncertainty and risk management
- Conflict resolution

In 1967, the US Department of Defense (DOD) introduced a set of 35 standards or criteria with which contractors must comply under cost or incentive contracts. The system of criteria is referred to as the *Cost and Schedule Control Systems Criteria* (C/SCSC). Many government agencies now require compliance with modified and updated versions of C/SCSC, whose purpose is to manage the risk of cost overrun to the government for major contracts. The system presents an integrated approach to cost and schedule management; and has been widely used in major project environments. It is intended to facilitate greater uniformity and provide advance warning about impending schedule or cost overruns.

The topics covered by C/SCSC include cost estimating and forecasting, budgeting, cost control, cost reporting, earned value analysis, resource allocation and management, and schedule adjustments. There is no doubt that the contemporary evolution of the Project Management Book of Knowledge (PMBOK) was influenced by the foundational contents of C/SCSC. The important link between all of these developments is the dynamism of the relationship between performance, time, and cost, as was alluded to in an earlier chapter of this book. Figure 10.19 illustrates an example of the dynamism that exists in cost–schedule–performance relationships. The relationships

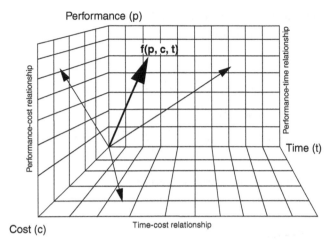

FIGURE 10.19   Cost–schedule–performance relationships.

represent a multiobjective problem. The resultant function, **f(p, c, t)** in Figure 10.19 represents a vector of decision, taking into account the relative nuances of project cost, schedule, and performance. Because performance, time, and cost objectives cannot be satisfied equally well, concessions or compromises need to be worked out in implementing C/SCSC or other project control criteria.

Another dimension of the performance–time–cost relationship is the US Air Force's R&M 2000 standard, which addresses the reliability and maintainability of systems. R&M 2000 is intended to integrate reliability and maintainability into the performance, cost, and schedule management for government contracts. Together, C/SCSC, R&M 2000, and other recent project control guides constitute an effective template for industrial project planning, organizing, and control.

To comply with C/SCSC, contractors must use standardized planning and control methods based on *earned value*. Earned value refers to the actual dollar value of work performed at a given point in time compared to planned cost for the work. This is different from the conventional approach of measuring actual versus planned, which is explicitly forbidden by C/SCSC. In the conventional approach, it is possible to misrepresent the actual content (or value) of the work accomplished. The work rate analysis technique can be useful in overcoming the deficiencies of the conventional approach. C/SCSC, by contrast, is developed on a work content basis using the following factors:

- The actual cost of work performed (ACWP), which is determined on the basis of the data from cost accounting and information systems
- The budgeted cost of work scheduled (BCWS) or baseline cost determined by the costs of scheduled accomplishments
- The budgeted cost of work performed (BCWP) or earned value, the actual work of effort completed as of a specific point in time

The following equations can be used to calculate cost and schedule variances for a work package at any point in time.

$$\text{Cost variance} = \text{BCWP} - \text{ACWP}$$

$$\text{Percent cost variance} = \left(\frac{\text{Cost variance}}{\text{BCWP}} \right) \cdot 100$$

$$\text{Schedule variance} = \text{BCWP} - \text{BCWS}$$

$$\text{Percent schedule variance} = \left(\frac{\text{Schedule variance}}{\text{BCWS}} \right) \cdot 100$$

$$\text{ACWP and remaining funds} = \text{Target cost (TC)}$$

$$\text{ACWP} + \text{cost to complete} = \text{Estimated cost at completion (EAC)}.$$

ACTIVITY-BASED COSTING

Activity-based costing (ABC) has emerged as an effective costing technique for industrial projects. The major motivation for ABC is that it offers an improved method to achieve enhancements in operational and strategic decisions. ABC offers a mechanism to allocate costs in direct proportion to the activities that are actually performed. This is an improvement over the traditional way of generically allocating costs to departments. It also improves the conventional approaches to allocating overhead costs.

The use of program evaluation review technique/critical path method, precedence diagramming, the critical resource diagramming method, and work breakdown structure (WBS) can facilitate the decomposition or breakdown of a task to provide information for ABC. Some of the potential impacts of ABC on a production line include the following:

- Identification and removal of unnecessary costs
- Identification of the cost impact of adding specific attributes to a product
- Indication of the incremental cost of improved quality
- Identification of the value-added points in a production process
- Inclusion of specific inventory carrying costs
- Provision of a basis for comparing production alternatives
- Ability to assess "what-if" scenarios for specific tasks

ABC is just one component of the overall activity-based management in an organization, and thus has its limitations, as well. Activity-based management involves a more global management approach to the planning and control of organizational endeavors. This requires consideration for product planning, resource allocation, productivity management, quality control, training, line balancing, value analysis, and a host of other organizational responsibilities. In the implementation of ABC, several issues must be considered:

- Level and availability of resources committed to developing activity-based information and cost

- Duration and level of effort needed to achieve ABC objectives
- Level of cost accuracy that can be achieved by ABC
- Ability to track activities based on ABC requirements
- Challenge of handling the volume of detailed information provided by ABC
- Sensitivity of the ABC system to changes in activity configuration.

REFERENCE

Badiru, Adedeji B, *Project Management in Manufacturing and High Technology Operations*, 2nd edition, John Wiley & Sons, New York, 1996.

11 Industrial Project Management Case Studies

AGV, Inc.

BACKGROUND

AGV, Inc., currently employs 55 people. The company was started in 1983 by three key people, Mr. White (company president), Mr. Nova (vice president marketing), and Mr. KC (vice president engineering). The charter of the company was to produce a highly specialized automated guided vehicle with a robotic arm mounted on board. This vehicle would be used in cleaning rooms, environmentally unsafe areas, and other specialized circumstances.

After the key members had developed the charter, Mr. White and Mr. KC went to work on financing. Mr. KC took his design and went to work finalizing the specifications. After financing was complete, they went to work on their first sale, and after 18 months, they had a firm order. The total system's price was approximately $1 million. Shortly after this sale, two additional sales went through. The first project required 18 months to complete. Many difficulties came up, but were quickly resolved. Considering the high level of technology used and the fact that the partners were new in that line of business, the project was successful. The system was purchased by a major US manufacturer. The second system sale was to the same US manufacturer. The third system was to a major manufacturer in Japan.

GROWTH PERIOD

During this time the company experienced major growth in size. It reached an employment level of 45 people (in 44 months). A setback was the termination of the director of software through mutual agreement. That post has since been filled and the new software director seems to be working out. The company also has had two additional rounds of financing. The majority of its financing comes from a major US financing company, with additional investment coming from a Dutch financing company.

CURRENT STATUS

The company currently has 65 employees. It has just taken orders on four additional systems, and work is going on to make major changes to two previous systems. To date, the company has completed only three systems (after 56 months). The marketing department is now working on 45 potential customers and expects orders to increase by 10%–30% in the next 9 months.

The fourth round of financing is underway. It appears to be going satisfactorily with the following exception. The US investors are very concerned that only three systems have been completed. They note that many schedules are running behind because of special features for individual customers. They also note that only a few key people in marketing know what the status of each project is, and that these marketing people are the only real contact the customers have with the company. A member of the investment team threatens with the following comment: "I expect to see some major organizational changes to occur before our next investors' meeting in 6 months."

CASE STUDY QUESTIONS

1. What is the real problem?
2. What should be done?

TRIPLE C IN INTERNATIONAL PROJECTS

Bakrie Pipe Industries is one of the major pipe industries in Indonesia. The current plant is located at Bekasi, a suburb area outside the capital city, Jakarta. With the increasing demands in pipe industries, the president of the company instructed the project director to construct and set up a new plant.

The project director was the person who held the highest responsibility for the success of the new plant setup. Based on the organization structure of the company, there were two divisions under his position:

1. The International Division, responsible for the out-of-Indonesia business
2. The Development Division, responsible for the coordination of all the departments that are involved in the project

The International Division of the project is located in the US, where its main tasks involve selecting and buying new machines, as well as ensuring good training programs for the utilization of the new machines. The headquarters of the company is located in Indonesia, where the project director is also located. So, the coordination, communication, and cooperation of the proposed project were difficult.

Owing to the complexities in coordination, communication, and cooperation between all the divisions and departments in making the project successful, the Triple C model was used extensively throughout the project cycle.

When the project started, the project director held the first general meeting with all the divisions and departments to inform them of the proposed new plant. He informed everyone of the project initiation and asked for each person's full participation when required by the project. As the International Division was not located in Indonesia, the communication was carried out by reporting directly though telephone and fax to the project director in Indonesia. The project director instructed the Development Division to call a meeting of personnel from the entire department required on this project to discuss further developments.

Cooperation from all the departments was crucial in achieving the deadline and success of the project. All the personnel of the departments that were to be involved had been instructed to give their full attention to the project. This full cooperation helped solve many problems and made the project a success by the deadline.

Coordination of the project was carried out by the Development Division, which was positioned under the project director. The Development Division was responsible for coordinating all the departments, assigning tasks to be carried out by each department, and identifying the critical tasks. This ensured that the most important tasks were performed on schedule.

ADVANTAGES OF TRIPLE C

The Triple C approach made this project a success. Many common communication problems were overcome or prevented. This was particularly important for the needed interactions between the International Division and the headquarters. The international tasks were integrated successfully, and personnel did not have coordination or communication problems because their tasks had been clearly defined and explained to them from the beginning. In sum, Triple C helped eliminate unnecessary costs.

The successful application of Triple C on this international project proved that the approach is effective and efficient. It has paved the way for further use of the approach in future projects of the company.

IMPLEMENTATION CONSIDERATIONS

There is often an initial "investment" cost of using Triple C. This involves the time, effort, and/or cost of setting up communication processes, particularly where traditional setups impede personnel communication. However, once the initial obstacles are overcome, the approach provides long-term benefits. However, a natural resistance to change may become a factor of concern in the attempt to implement Triple C. Personnel have to be convinced to reorient their task priorities to attend to the requirements of the new project. A lot of time may have to be invested in the attempt to secure the commitment of some departments to the proposed project, and many communication efforts may have to be made before full cooperation is achieved. These initial obstacles should not, however, discourage the project efforts.

INTERCONTINENTAL PROJECT COORDINATION

THE PROJECT

The project in this case study comprises of the engineering, procurement, and construction of seven liquid gas tanks together with their ancillary system and control building. The use and capacity of the tanks are as follows:

1. Three liquefied natural gas (LNG) tanks of 80,000 m³ each
2. Two liquefied petroleum gas (LPG) propane tanks of 50,000 m³ each
3. Two LPG butane tanks of 50,000 m³ each

The ancillary systems which are part of the project include:

1. A propane and butane vapor recovery systems
2. A low-pressure flare system
3. An offsite control room

This project is to be built on DAX Island in the Persian Gulf (Middle East), approximately 300 mi from the city of Abu Dhabi (the capital of the United Arab Emirates). DAX is a small island of 2 mi^2, and has a population of 5000, all of whom work in the only two plants on the island. Owing to the large project size, the island will have an additional 3000 people working and living on the island during the construction phase, which is a considerable increase in the island population. The island has salty soil and cannot grow any vegetation. Also, there is no source for freshwater except for desalinated seawater; because desalination is a very slow process, it may become the cause of considerable delays when it comes to concrete pouring.

Each of the seven tanks will consist of two separate and structurally independent liquid containers—a primary inner metallic container and a secondary outer concrete container. Each of the containers will be constructed out of material suitable for the low-temperature liquid.

Each container should be capable of holding the required volume of stored liquid for an indefinite period without any deterioration of the container or its surroundings. The secondary concrete container should be capable of withstanding the effect of fire exposure from the adjacent tanks and an external impact of 21 tons traveling at high speed without loss of structural integrity. Adequate insulation should be provided between the primary and secondary tanks to limit the heat in-leak. All the tanks should be provided with the necessary pumping and piping systems for the receipt, storage, and loading of the LNG and PLG from the nearby plant to the different tankers. A blast-proof offsite control room should be provided. All storage and loading operations should be controlled from this control room. The project total cost is estimated at US$600,000,000. The detailed project critical path method (CPM) consists of around 48,000 activities.

ORGANIZATIONAL PRINCIPLES

GASC, the company awarded this project, is one of several operating companies, a majority of whose shares are owned by NOCC, the national oil company of the Emirate of Abu Dhabi. With main offices in Abu Dhabi city, NOCC is wholly owned by the government of Abu Dhabi and plays an extremely important role in the national economy. GASC is an operating company that had never managed any engineering or construction project. GASC turned to its mother company, NOCC, for help and signed an agreement with NOCC for the management of this project.

NOCC is headed by a general manager responsible for the eight different directorates, one of which is the Projects Directorate. One of the divisions of the Projects Directorate is the Gas Projects Division, headed by a manager. The Gas Projects Division has been selected to manage the above project.

NOCC's share in GASC is 60%, and the remaining shares are owned by TOTAC of France, SHELC of Holland, PBC of Great Britain, and MITSC of Japan. While it is considered convenient to have the major owning company manage the project for GASC, such an arrangement has its own drawbacks, mainly due to the fact that the client company (which should have the final say in its own project) is a subsidiary of the hired management company.

The Gas Projects Division is staffed with a skeleton staff. Although very capable, this staff can only oversee the management of the project, but cannot perform all the actual engineering, procurement, and construction management, and so Texas-based KELLC has been selected to provide those services under the direct supervision of the Gas Projects Division. KELLC's scope of work includes the basic design of the tanks, the detailed design of all the piping and ancillary systems, the procurement of all free issue materials, and the management of construction. All the engineering and procurement activities are to be performed out of the KELLC regional office in London, England. KELLC, a well-known process engineering firm, has limited experience with concrete tanks: their selection was conditional on their agreement to hire the Belgian civil engineering firm TRACT as a consultant to help them in the critical civil engineering problems. NOCC will also hire the America-based consulting firm DMRC to do the soil investigation and testing work.

The construction work is packaged into 15 small contracts and 1 large contract (75% of the total construction work). All the small contracts were awarded to local construction companies, while the main large contract, which included the tanks, piping and ancillary systems, was awarded to Chicago-based CBAIC. CBAIC will have to open three new offices: one office in London, next to the KELLC regional office, during the engineering phase; another office in Abu Dhabi city for the construction management; and a third office on DAX Island for the construction operations. CBAIC is a reputable tankage contractor. However, its experience in concrete tanks is quite limited: They will have to hire the French civil engineering firm SBC as subcontractors. SBC's experience in low-temperature concrete is limited; they will have to hire the specialized Belgian firm CBC as a consultant.

The project's safety requirements are very high: The safety of those living on such a small island in case of any accident is a major concern to the owning company. To ensure that the required quality and safety standards are achieved, NOCC will hire the French "third-party inspection" company BVC as a consultant.

Since the engineering office is in London, the French and the Belgian engineers are to commute to London as necessary to provide their input to the project: This will continue for the whole engineering phase, which is expected to last 2 years.

Procurement activities will be handled out of KELLC's London office. Materials and equipment are to be delivered to the construction site. Supplies will be purchased in the open market at the most competitive prices. Steel will be purchased from Japan and Belgium; pipes from Germany, France, and Japan; valves from Sweden and France; pumps from the US; compressors from Switzerland and Japan; and vessels from Italy. A total of around 600 purchase orders will be issued.

ORGANIZATIONAL SETUP

The organizational setup has to be quite flexible and able to change in accordance with the project requirements. As it is expected to be a fast-track project, the most active period will be the second half of the engineering phase, which corresponds to the first half of the construction phase. The organizational setup for the project stretches across national boundaries, practices, and regulations. It is, thus, very important that efforts of the project staff be closely coordinated.

Case Study Questions

1. As this is a project that crosses national boundaries and time zones, communications will without doubt be a major problem. It will be desirable to determine a time frame on any given day when all project sites can communicate over the telephone. To address that problem, determine a time window on any given day (24 hr) that all the project offices in the following cities or countries can communicate by teleconference: London, Rome, Chicago, Abu Dhabi, Tokyo, Geneva, Bonn, Paris, Sweden, and Belgium.
2. Carefully go through the narrative of how this project is organized and then develop an organizational chart that conveys that organizational setup. The chart should show the various company names, their responsibilities, their interrelationships, and their locations.

SQUEEZING THE EAGLE

BACKGROUND

American Peripheral Company (APC) is a manufacturer of computer peripheral equipment. APC had dominated their market until recently when Japanese manufacturers acquired the technology to compete effectively with APC. Price erosion forced APC to concentrate on high-performance products, which, however, had the lowest volume; the result was decreased revenue. Combined with a general industry slump, this has forced APC to sell part of the company and lay off 40% of its employees.

APC implemented project management 2 years ago to boost productivity and shorten product development cycles. This has been only partially successful because project managers develop schedules with inputs from line managers, only to have upper management shorten them by 20%–40% to meet perceived market windows. Project managers find that they are 4–6 months behind schedule when the project begins. Most of APC's employees have been with APC for 10–20 years. Though morale is low, many engineers work an extra 5–10 hr per week without pay in an attempt to meet their deadlines.

THE PRESENT

Bob Stephens is the project manager for the Eagle 3, which is currently under development. The project is 11 months into a 15-month schedule and is 2 months late. The first customer test units are due to be delivered January 2

Bob has just left a meeting with the general manager and the new products manager (Bob's boss). Bob was told that a major customer must have four test units by Thanksgiving to consider the Eagle 3 for use in his new computer. As on previous occasions with previous projects, Bob was told that if APC gets this contract, the company will "get well." The general manager also told Bob that there were no engineers available to reassign to the Eagle 3 project, and that the budget would not allow for overtime pay.

1. How should Bob motivate his project team?
2. What options are available for meeting the Thanksgiving deadline?

GENERAL COMPUTER

General Computer is a large-scale manufacturer of state-of-the-art computer equipment, with sales over $1 billion annually. Their largest factory, in central Arkansas, employs approximately 4000 people. This factory currently manufactures the 500 ST series of business computers.

In 1977, when the factory started manufacturing the 410 series of computers, the managerial organization was reworked to add a matrix structure project management group to the staff. It consisted of one manager, one assistant manager, and seven technical/engineering personnel. This group made marked improvements in the manufacturing process in the four years from 1977 through 1981.

In 1981, the 410 series was phased out, and the new 510 ST series was introduced. The 510 ST series was much larger in manufacturing scope and was more complex than the old 410 series.

Because of these factors, management decided to add to the project management staff. Instead of adding to the existing project management group, management added another department with another eight nontechnical people. The project management functions were then split, with the old group overseeing the engineering/planning functions and the new group taking the tracking/implementation functions. Both groups reported to the same manager but to different assistant managers.

After an initial adjustment period, both groups performed well in their respective jobs. After 6 months, however, the quality of the work began to suffer. Certain jobs were not being completed, the excuse being that the offending group thought the other group was supposed to do it. Some jobs were being duplicated, often with conflicting conclusions. Bickering between the groups became more frequent. More and more functional groups began to complain to the project manager. Finally, the project manager called a meeting of both groups. He started the meeting by asking, "Why are we having these problems?"

CASE STUDY QUESTIONS

Assume you are a consultant sitting in at the meeting.

1. What would you expect to hear?
2. Do you feel that the new matrix structure was properly implemented?
3. What are your recommendations?

ADVANCED COMMUNICATIONS COMPANY

Advanced Communications is a multinational electronic components and manufacturing corporation. The microwave division (ACM) has a manufacturing plant in Wichita, Kansas. Bill Howard is a failure-mode-analysis engineer who performs testing on components for ACM's biggest product, the Wave-16. The Wave-16 is a short-range communicator used extensively in the military. Currently, ACM has a $400 million contract to produce Wave-16s. The contract will expire at the end of the next year with almost certain renewal if no problems are exhibited by ACM's product.

Four months ago Bill was approached by a product engineer, Al Fleming, about a component that was causing problems both in the factory and in the field. This component, the RB175, is a complex integrated circuit that made up the core of the Wave-16. As Al became more familiar with his product, his testing of the RB175 became more precise. Lately, Al had noticed a trend in the factory of excessive RB175 failures. As the RB175 was an expensive custom component made at another Advanced communications facility, Al had collected all the failures and retested them. During retest, 85% of the failed RB175s passed. He tested them again, and this time only 35% of the failed RB175s passed. He decided that he needed the equipment and 15 years of experience that Bill could provide. So, he sought Bill's participation.

After 2 weeks of intense testing under all conditions, Bill and Al came to the conclusion that specific temperature changes and humidity levels caused a specific circuit in the RB175 to react unpredictably. Both their direct superiors were then informed of the problem. Considering the importance and high cost of the RB175 ($4500 per unit), a task force was formed to solve the problem. Forming the task force alone took 2 weeks because of low interest at the RB175 component-manufacturing facility. This facility, Advanced Communications Components (ACC)–Denver had never observed this phenomenon and felt that ACM was overstepping its boundaries by performing other than purely functional tests. Cooperation was only received after ACM's plant manager contacted ACC's plant manager.

Once the task force was formed, ACC would not cooperate until it had repeated all the tests Al and Bill had previously performed and documented. This took four more weeks because of poor staffing and a general feeling that a problem did not exist. Once the failure was validated, ACC became cooperative but ignored complaints of delays. ACC also informed the task force that the circuit that was failing was indeed packaged at the Denver facility but generated at ACC's Los Angeles facility. Another 4 weeks were needed to bring the ACC–LA group into the task force and return to the point where Al and Bill had been several months previously.

Two months later, a solution was yet to be found, and so all the research and development (R&D) people who had been involved in the RB175 were brought on board. Both Al and Bill had by now been assigned full-time to the task force and were relieved of all other duties. Bill's lab equipment was dedicated to this project only. Meanwhile, the Wave-16 was still being produced on schedule with the hopes that minimal field failures would occur. The customer was not informed of the problem as a new contract was coming up. Any field failures were replaced immediately at no cost to the customer, but a projected loss of $7000 per replacement in labor and parts was incurred by the ACM task force. One year later, it was decided to redesign the RB175 and the Wave-16 for less cost, taking advantage of new technology. Advanced Communications presented the new product design, the Superwave-20 to the military as a better product and was awarded the contract. Al and Bill were both transferred to R&D, where they could incorporate their design ideas into the new RB180.

1. What management reorganization may be needed to avoid problems of the past in the new contract?
2. What employee communication problems can you identify in the organization of Advanced Communications Company?

METROPLEX MANAGEMENT

"What am I going to do," Tom, a project manager, asked Dr. Packer, head of the Applied Research Division of Metroplex International. "All the line people assigned to my project are slowing down production and doing other things to cause my project to come in late and way over budget. They are making it necessary for me to pay them for overtime to get a job done that they should not have had any trouble whatsoever getting done in the time that I scheduled for it. The customer is calling me daily wanting a progress report on the project. I am getting tired of telling him that we are behind but are trying to get the project back on schedule."

Tom Reese is a new project manager on a large $50 million project for Metroplex, which is a $220-million-a-year firm. This contract amounts to approximately one-fourth of the firm's yearly business, and if the project goes well, there is a promise of follow-up business of increasing values. Tom has been hired for the job of project manager specifically for this project because he has a reputation of being a very good project manager who gets the job done on schedule, under budget, and with the highest of quality. The executives of the company felt that they needed someone of this caliber for this job. This action has, however, angered many of the employees because the historical policy of the company has been one of promoting people from within.

Many of the employees feel that Mike Johnson, the present line manager, should have gotten the new project manager's job. Therefore, the line people are doing substandard work on the project deliberately to ruin Tom's reputation.

"I need your help to get through to Mike Johnson and his people that this is a very important project for the company. We must bring this project in on schedule and not overrun the cost, either. They need to understand that they are just hurting themselves by slowing down this project," said Tom.

"Well," said Dr. Packer, "that is your problem. You are the high-powered project manager on this project. You are supposed to be able to work with all the people that you need to handle your project. Isn't that the job of the project manager? You solve the problem."

"Yes, that is the job of the project manager," said Tom, "and I have tried, but Mike Johnson won't listen to me. He says that I have no authority over him and that you, our boss, will have to tell him to do things differently. In the meantime he just keeps right on allowing his people to waste time and money by causing delays in my project that are making the customer very unhappy. If we don't do well on this contract, then our customer may not award us the follow-on contracts that are being proposed for this project, and they will be lucrative."

"He is absolutely right about the authority level," said Dr. Packer, "but I am still leaving it up to you to get this problem solved; if we lose the follow-on contracts, then heads will roll. Do you understand?"

1. Is promotion from within a company a good policy? Why or why not?
2. Is the claim about sabotaging Tom's reputation accurate?
3. Identify some of the potential problems associated with putting personal interests above company goals.
4. Why is Dr. Packer not cooperating in solving the project problems?
5. Is power play a major factor in this company?

TECH SOUND BANKRUPTCY

Tech sound, Inc., was a manufacturer of sound equipment for over 50 years. Its product lines included electronic amplifiers and various acoustic speakers. The brand name "TECH 2000" once was the leading brand name, and considered to be the Cadillac of the speaker industry.

Tech Sound had two manufacturing facilities, one located in California and the other in Oklahoma. The Oklahoma factory production volume was responsible for over 70% of the sales even though the corporate headquarters was in California. The product lines covered industrial and professional speaker units, home stereos, car stereos, and outdoor speakers. All but the industrial and professional speakers were made in Oklahoma. The general techniques and concepts of designing a speaker had not changed for over 30 years.

The components needed to assemble a speaker can be divided into two categories:

1. Hardware such as the frame, the top plate, the bottom plate, a pole piece, screws or rivets, and housing (for outdoor speakers only)
2. Software such as voice coils, spider/diaphragm, corns, corn caps, and gaskets

These two types of components were manufactured by Tech Sound itself because of the close and precise tolerance requirements. Therefore, Sound Tech had large departments devoted to machine shop work and coil winding in both factories.

The sound equipment market is a very competitive one. Products from the US, Japan, and other countries are all trying to gain bigger market shares. Sound Tech had been experiencing losses in revenue for the last several years, and it had an outstanding debenture that would mature in another 2 years. Top management felt a regrouping of the financial structure and a major cost reduction were necessary to keep the company from going under. After several meetings, the decision was made to close the factory in California and move the headquarters to Oklahoma. The property in California was sold to pay for the debenture. To have enough floor space for the incoming industrial and professional speaker line and to reduce cost, the final assembly process (assembly of software and hardware to a complete speaker) of the outdoor speakers were moved from Oklahoma to Mexico. The steps required to implement the above decisions are here presented, with the responsible departments listed in parenthesis:

1. Planning
   a. Generation of equipment and machinery list (engineering and production)
   b. Production scheduling (production control)
   c. Generation of material inventory list (production control)
   d. Drawing up of time schedules (project manager)
   e. Transportation (project manager)
   f. Layout of new factory floor (engineering)
   g. Documentation transfer (all)

2. Physical movement
   a. Shipping to new location (project, engineering, contractor)
   b. In-house movement of equipment and machinery (Project engineering, engineering, and contractors)

3. Training and start-up in new location
   a. Training of assembly personnel
   b. Training on quality control
   c. Training on engineering
   d. Review and evaluation

Tech Sound completed its manufacturing facilities relocation project in 20 months. The production continued at its new location with some problems, mainly caused by inexperienced workers (no employees were willing to be relocated to the new plant). After the learning period was over, the level of production quality and performance went back up to the previous levels. However, 1 year after the project completion, Sound Tech filed for bankruptcy, and eventually sold out to another company.

CASE STUDY QUESTIONS

1. What were the underlying causes of the problems of Sound Tech Company?
2. Were the problems due to poor organizational setups? Should things have been done differently? What would you recommend for a company in an identical situation?

NEW PRODUCT DEVELOPMENT PROGRAM

PROJECT BACKGROUND AND PLAN

The increased demand for widgets in this country and in international markets has provided the impetus for Widgety Manufacturing Company to create a new manufacturing facility to produce and distribute widgets. The current project involves the construction of the production facility. The completion of the project will require the following:

- Capital
- Industrial-grade construction materials and some subcontracting

- Manufacturing and plant maintenance equipment
- Personnel
  - Nontechnical labor
  - Technical labor
  - Engineering support
  - Managerial and administrative staffing

The plant location should be chosen to facilitate distribution of the product to international markets. This entails finding a location with good access to major shipping, railroad, and highway transportation routes. The production of widgets will meet an increasing international demand for the product, which is forecast to remain high for some time. This will benefit not only the company, but should also improve the country's balance of trade in the international marketplace. The project will require an initial investment of:

- $5 million, which will be used as detailed subsequently
- $3 million for materials
- $1 million for land acquisition and fees
- $1 million for personnel expenses

The goal of the company is to complete the facility within 18 months of the acceptance of a bid. The three main objectives within the overall project goal are as follows:

- Raising of capital for the project
- Acquisition of land and building permits and adherence to local regulations and tax structure
- Plant construction and equipment installation

The capital is to be raised through sales of stock and loan procurement. Clearly, the achievement of this objective will be marked by the production of adequate funds for the project. The acquisition of land for the plant also entails choosing a location, which has a set of requirements regarding local building codes, zoning laws, tax structure, and environmental regulations compatible with the existence of the plant. This objective will be deemed complete when a signed agreement with a community has been reached, when all relevant building permits have been acquired, and when appropriate tax credits and costs have been addressed.

The third objective is the actual construction of the plant and installation of the industrial machinery. The success of this objective will be measured by the operational ability of the plant, specifically, the ability of the facility to meet all applicable building codes, safety codes, and environmental regulations, as well as the ability of the assembly lines and environmental controls (e.g., heat, air-conditioning, plumbing, electricity) to function properly.

PROJECT APPROACH

Managerial: The managerial approach for the overall project will be the "management-by-objective" approach. However, in identified subtasks, local task heads may have latitude in using the management-by-exception approach, if this is deemed more efficient, particularly from the point of view of cost control. In these cases, the local task managers must provide a reasonable, documented rationale for the decision. Accountability will be determined as part of the managerial hierarchy, as well as being based on scheduling concerns, where completion of one task is necessary for completion of others.

Technical: The technical approach to be used in this project relies partly on production technology used in many industries, as well as on newer technology, which should improve plant efficiency. The use of newer technology requires that the company identify appropriate personnel for training in the new field in order to facilitate integration of the technology into the plant.

POLICIES AND PROCEDURES

Policy: Policy during project execution is divided into two categories: financial policy and physical implementation. Any function requiring the expenditure of funds requires the approval of the manager assigned to the functional group in which the expenditure will take place. Any expenditure exceeding $Z will require the combined approval of the project manager and the manager assigned to fiscal accounting. Physical implementation is a broad category comprising most other policies and their related functions within the project. These policies include setting working hours, length of breaks, worker attire, managerial responsibilities, methods for hiring and terminating, sick leave, and site safety. This list is not exhaustive and will be reassessed and adjusted on a regular basis.

Procedures: Each functional group will have a predetermined budget. Any expenditure within this budget requires written authorization by the group manager prior to disbursement. Any cost overruns must be approved by the project manager after a meeting between the manager, fiscal authority, and group manager.

The policies regarding personnel and site safety will be implemented at the group-manager level. The group manager will ensure that the employees work a full 8 hr day, and that sick leave is only granted when notification is given within 8 hr of start of shift. Site safety will be based on current OSHA regulations.

CONTRACTUAL REQUIREMENTS

The communication structure starts with the weekly project review meeting to be attended by all group managers and the project manager. Results of this meeting will be transmitted by the group managers to their respective groups. Each group will have a daily meeting in the morning for subforemen to address local issues with the group manager. It will be the group manager's job to ensure that significant project concerns from local workers are brought to the attention of the project manager, and that universal policies and procedures determined at the upper management levels will be disseminated throughout the job site.

Performance specifications will be set, by the project manager, in the macro setting, and by group managers at the microlevel of management. Performance specifications will include factors such as meeting deadlines, meeting architectural requirements, and meeting cost. Review of performance by each group will be a part of the weekly management meeting.

248   Industrial Project Management: Concepts, Tools, and Techniques

During a project, large amounts of data are collected while analyzing and reviewing the progress of the project. This data will be kept locally by each group manager, as well as being sent to Management Information Service for overall collation and correlation.

PROJECT SCHEDULE

The first step in the project is procurement of funds. This will be accomplished within 45 days. When 10% of the desired funding is achieved, preliminary architectural drawings and engineering analyses will be performed. If within 45 days, sufficient funding is not obtained, the project will be terminated until such a time as it is deemed more financially feasible.

The second phase entails land acquisition. An appropriate site that meets the cost, location, and local environmental criteria will be chosen. This should be accomplished within 30 days. Completion will be defined as that time when all legal, contractual papers have been signed, fund and property rights transferred, and all relevant building permits acquired.

As soon as a site has been identified, and during the time the paperwork for acquisition is moving forward, a construction firm will be identified. Additionally, engineering support personnel will be recruited. The construction firm will be responsible for hiring both the nontechnical and the technical personnel. This portion of the project can run partially concurrently with the land acquisition, but must be completed within 15 days of the finalization of the land deal. The construction phase of the project will last the remaining 13 months. This phase can be subdivided into several phases:

1. Land preparation (15 days)
2. Necessary in-ground and electric (45 days)
3. Foundation (30 days)
4. Skeletal structure (60 days)
5. Roofing (30 days)
6. Above-ground wiring, plumbing, and ducting (60 days)
7. Large equipment installation (60 days)
8. Enclosure (45 days)
9. Remaining equipment installation (30 days)
10. Testing of plant functionality (15 days)

The schedule is designed for sequential completion. However, concurrent work between steps 2 and 3; 5 and 6; and 7, 8, and 9 will reduce the overall time. The sequential schedule builds in tolerance for potential delays. Each phase listed here will have a project schedule generated by the functional group manager for that phase.

RESOURCE REQUIREMENTS

The estimated cost of the project is $5 million, and $3 million is to be spent on materials as follows:

Foundations: $300,000
Electrical work: $500,000

Plumbing, ducting: $500,000
Structural material: $1,000,000
Equipment: $1,000,000

Personnel expenses will be broken down as follows:

Nontechnical labor: $400,000
Technical labor: $200,000
Engineering support: $150,000
Management: $250,000.

Part of the $150,000 for engineering support will be used for training specified engineering personnel on the new process equipment and design to be implemented in the factory.

Performance Measures

The basic measures of performance will consist of the following:

- Completion of task on schedule
- Completion of task at or below budget
- Completed task meeting or exceeding all specifications

A weekly audit of expenses for each task will be performed to monitor progress. Completion of the task on schedule is easily verified, but weekly progress reports will be mandatory so as to enable adjustment of scheduling for other tasks if needed. Each task will also be divided into distinct objectives. When each objective is met, an assessment of whether the completed phase meets the specifications of the phase will be made. If the specifications are not met, corrective action will be taken before proceeding. Dividing each task into subtasks should be as compact as possible to ensure that excessive amounts of work are not completed before the identification of subpar components.

Contingency Plans

During construction, funds exceeding 15% of the estimated cost will be sought and procured. This will eliminate any delays due to cost overruns. Insurance will be purchased for major natural disasters, as well as liability, and indemnity. Alternate scheduling will be constantly re-evaluated to account for delays caused by weather and supply inadequacies.

Tracking, Reporting, and Auditing

The project manager will be responsible for being knowledgeable about the progress and state of each subgroup. Each week, group managers will supply the project manager with a synopsis report of the current states of his/her group. Each group manager will also submit a more in-depth analysis of the group's operation, in terms of budget, progress, and performance achievements, to the management information service. This division will collate and correlate all data and will be able to apprise the project manager of the overall state of the project at any time.

CASE STUDY DISCUSSION QUESTIONS

1. Which organization structure should be used for this product development project? Draw a sketch of the structure.
2. What are some of the international market considerations for this project?
3. Discuss the potential problems with the sick-leave policy.
4. Is the 18-month project schedule realistic? Why or why not?
5. Outline the other considerations, if any, which should be included in the project's plan.

FASTRITE MANUFACTURING COMPANY

In the FASTRITE Manufacturing Company (FMC), the engineers are required by the line managers to provide the assembly personnel with manufacturing layouts for every item they manufacture. These manufacturing layouts contain all of the information that the shop needs to build the product. The problem with these documents is the fact that they are always subject to changes due to several factors. Currently, the process of issuing the manufacturing layout takes approximately 38 days. The line manager (Diana) feels that this long lead-time is not acceptable. Therefore, she called the engineering manager to complain. The engineering manager (Bill) understood the problem and agreed to help.

Bill decided that he would ask John, one of his engineers, to work on this problem. Bill called John into his office to explain to him the situation. Bill said that he was aware of the problem and that he knew why the problem existed. Bill was pleased with the quick response and wanted to know what the problem was and how he was going to resolve it. John proceeded to explain that the current procedures are uncontrolled and do not contain time constraints: "The process that is associated with issuing manufacturing layouts is far from ideal. Everybody wants to influence the layout in one way or another. In order to complete a manufacturing layout, you are required to obtain nine other engineers' signatures. Each of these nine engineers has a functional interest in the layout. The problem is that there are no time constraints that indicate how long after you receive a layout you should supply your input and pass it on."

Bill then understood what was happening. The layouts were sitting idle on engineers' desks for days before the engineers would do their part. "How are you going to resolve this problem?" Bill asked John. John thought for a minute and decided to develop a program evaluation review technique (PERT) chart for the operation. Then he went to each of the nine functional engineers individually to determine the order of precedence and the duration of their activity. After compiling his data, he drew the PERT chart to determine the overall expected duration. He was pleased to find out that the expected duration was only 12 days.

John's next dilemma was what to do with this information in order to decrease the current duration. He decided that the best way was to publish his findings to the functional engineering managers. In doing this he could let the managers convey the information to their engineers, along with instructions to adhere to the time schedule. He reported back to his boss to explain his proposal and get his approval. Bill reviewed the proposal, was thoroughly impressed with the concept, and gave John his approval. John's concept worked. The duration was shortened to an average of 11 days. However, John offended some people by giving the engineering managers the impression that their functional engineers were not doing their jobs expeditiously.

Case Study Discussion Questions

1. How could John have introduced the new time constraints to the functional engineers without offending them?
2. Should Bill be blamed for the animosity that developed?
3. As the concept worked to the benefit of the organization, should anyone be concerned that some people did not like the way the concept was introduced?

PRODUCT DEVELOPMENT RACE: KRACON ENTERPRISES VERSUS UNJE COMPANY

Kracon Company and Unje Enterprises are engaged in a competitive time-to-market race for a new product. Both companies are producers of fine chocolate. The R&D departments of both companies have been working for the last 6months on a new chocolate formula that is a rich and creamer chocolate for milk chocolate lovers. Currently Kracon is the industry leader with Unje a close second. It just so happens that both R&D groups have come up with two different types of technology for their new chocolate formulas. Each company, through the grapevine, has learned that the other company is coming out with a new product similar to their own. If the launching of the new products is successful it would mean a higher margin and lower cost product. Now the race is on to see who can be the first to market.

Unje Project

Culture

The Unje Company has no formal structure for its engineering standards, procedures, or quality guidelines. The company has been completing projects using partial guidelines and previous experiences of senior engineers. Forms are developed on an as-needed basis; standards and specifications are developed for one project and passed from one engineer to another to be reused. There is no formal training program for engineers; rather, the company believes that field experience is the best learning forum. There is limited communication between the different departments and very little collaboration. The company's senior management has realized that in order to become number one in the industry, the company as a whole has to change. Unje had already had four high-profile projects that were unsuccessful due to poor planning and design. For this reason, various initiatives have been set in motion. The new product is a direct result of a margin growth initiative and considered one of the success stories for R&D and marketing. There is a lot of focus on the successful execution of the new process line. The senior management at Unje made it known that the objective is to execute quickly so as to launch ahead of Kracon. The new process relies on new equipment which can only be purchased from one vendor and relocation of existing equipment.

Unje's Race Market: "I don't have to time to tell you what I want"

Once R&D established that the pilot system was successful and they were ready to move forward with the full-scale model, engineering was brought into the discussion. The shutdown for execution was set for the end of the year. Engineering was given two tasks: cost estimate and capital request completion. Detailed design was not given a priority because of time constraints. R&D was ready to move forward in June and senior management expected a cost estimate within one month. Engineering moved forward with cost estimate based on a limited design and a limited scope and a contingency of 10%. However, during the operational analysis, errors in the scale-up from pilot to full production were discovered. Owing to time constraints and the high-profile nature of the project, the concerns were under played and the cost estimate was completed.

When the cost estimate was presented to senior management, it was promptly rejected. Senior management stated that the budget needed to be reduced by 20% and the operational errors resolved; however, the time frame for execution needed to remain the same. By the time engineering and R&D had resolved the operational errors and reduced the budget by 20%, it was August. The project was sent to all the necessary managers for approval, which was obtained in September. During the revision stage of the capital request, negotiations with the sole vendor were ongoing. Unje pressured the vendor for a shorter delivery time and a reduced purchase price. The vendor eventually agreed to the shorter delivery but would not reduce the equipment price. Unje and the vendor signed an agreement in late September.

Engineering assigned three internal engineers to the project and sent out an RFP for engineering services to Mudt Engineering. Mudt responded that the time frame to complete design was inadequate and they could not complete the job unless the job was separated into critical and noncritical items. Unje agreed and the critical items were determined to be needed 2 months prior to execution and the noncritical 1 month prior to execution. Mudt began final design engineering and found several gaps in the original scope of the project as outlined in the capital request: the utility requirements were more than what is available in the plant and there were several infrastructure issues. Every time an issue came up, it took time to resolve because there were no standards or specifications and no clear point-of-contact person. Mudt was running out of time to complete the design and Unje was pushing to have everything done. In the end Mudt completed the design and a bid package was sent out to local contractors. The bids were received and the contractors selected. Internally Unje was obtaining estimates for peripheral equipment and other services and finding that the cost estimate submitted was 40% below the actual cost to complete the project based on the final design. Unje had to return to senior management and complete a supplemental to the capital request before the project was even near execution.

The selected construction contractors expressed concern with the time constraint for completion and the lack of accurate drawings and design completion. In fact, there were several issues during the construction phase. Equipment was located incorrectly, utilities were routed incorrectly (in some cases not at all), existing equipment was not shown correctly, the plant layout was not updated, and the contractor was continuously running into unforeseen utility lines. When Unje took issue with Mudt for the low-level engineering, they were told that Mudt did the best they could in the time they were given. Each day the scope of the project grew as more gaps were discovered in the design and requirements of the project. The contractor had to work 24 hr a day to complete construction on time and was 130% over budget.

In the end, Unje started up the new line at the beginning of the year but only at 10% of the predicted production rate and below quality expectations. The operators were being trained during startup because no time was scheduled for training prior to starting up. By the time Unje had worked out all the variability in the new line and was able to make saleable product, it was late March.

KRACON ENTERPRISES

Culture

Kracon over the 5 years has developed what they call EPMG, or Engineering Project Management Guidelines. The EPMG is a formal outline of the steps for a project life cycle. The EPMG is one step of the Kracon commercialization process and engineering has ownership of EPMG for every project. Engineering is required to submit an EPMG booklet (EPMG-P) that consists of an executive summary, cost analysis, project realization and preliminary scope analysis, core team definition, roles and responsibilities, contract procurement plan, equipment procurement plan, detailed schedule, quality sign-off sheet, process description, preliminary design drawings, and risk and opportunities. The EPMG-P is given to the business leader, who then writes the capital request based on the information in the EPMG-P. If the capital request is approved, the signed EPMG-P is sent back to engineering for execution; at which point an EPMG-E is completed. The EPMG-E is the finalized version of the EPMG-P with a detailed scope analysis, finalized contract and equipment procurement schedule, finalized schedule, risk mitigation plan, and finalized design drawings. All new-hire engineers are assigned required courses that must be taken within the first year of employment. These classes include time management, quick thinking, chocolate 101, marketing for engineers, and internal project management training. Kracon has made it a practice that the EPMG-P must be completed for all submitted projects; projects submitted without an EPMG-P are rejected and sent back to the business leader. Kracon, like Kracon, has many initiatives that it has started and the new chocolate process is a direct result of margin growth and value-add initiative. Senior leadership wants to get to market ahead of Unje Enterprises, but they also want to maintain the standard of quality to their customers. Kracon places quality at the top of its deliverables list.

Kracon's Race to Market: "We are the tortoise"

In June, R&D informs the engineer that they are ready to move forward from the pilot system to full scale product. Engineering and R&D get together to form a core team of five people and one project manager. A kick-off meeting is scheduled at the plant with the core team members. During the kick-off meeting, the discussions get heated. Quality is not satisfied with the product that R&D has been producing and wants R&D to complete some more changes. Procurement is concerned that the product can only be manufactured on a specialized piece of equipment that has to be purchased from sole source. The core team cannot agree to move forward with an EPMG-P, and installation is scheduled for the end of the year. The project manager agrees that if the product is not meeting existing quality standards, then the project cannot move forward. R&D is ordered to return to the lab and make changes to get the product within specifications, while engineering concurrently contacts and an engineering firm is asked to begin preliminary engineering design and estimation. The project manager computes the schedule and determines that execution cannot occur at the end of the year but rather in mid-February. This information is communicated to management, who is not happy with the change in plans, until the issue of quality is raised. In the end management states that construction cannot occur any later than late January or Kracon runs the risk of being late to market. The core team meets again in late June and determines the necessary steps to begin execution in late January. Key milestones are determined and resources are assigned. R&D agrees to assign more resources and to go outside to help find resolution to the quality issues, and engineering agrees to move forward with the EPMG-P with a higher contingency level in certain sections and procurement begins negotiations and performance agreement talks with the equipment vendor. An RFP is sent to an engineering firm for assistance with design and drawings for the EPMG-P.

In late August, the core team has a final review of the EPMG-P and R&D's resolutions to the quality issues. The core team is in agreement on the cost and the preliminary design. Procurement has come to an agreement with the vendor, which includes additional time to build the equipment but at a lower cost. The vendor can meet the revised schedule time line. The EPMG-P is sent to the business leader, who completes the capital request. The capital request is sent back because senior management feels the capital cost is too high and wants to review the numbers. The capital request is revised and resubmitted at a cost estimate lower by 10%. The capital request is approved in mid-September. Engineering contacts the same engineering firm and requests a proposal for final design services and at the same time, engineering contacts a construction management firm for a proposal for construction management services. Both contracts are awarded. Review meetings (at 30%, 60%, and 90% points) are held with the core team and key plant personnel during the design phase. The final design is completed in late November and a request for quotation complete with drawings and detailed scope is issued to contractors for bid. The construction bid is awarded in early December and construction work is started in mid-January.

At this point the work, although behind the original schedule, is ahead of the revised schedule. During construction several issues come up with missed utility tie-ins and unforeseen infrastructure concerns. However, because time was taken to update the existing plant drawings, none of the issues cause major delays to the completion of construction. The project manager actually finds that he is 3% below the submitted budget in the EPMG-P. Construction is completed in late February and a walk-through with quality is performed. Operators are brought in a week prior to startup and trained on the new equipment. The process starts up at 25% efficiency the first week, which is within the startup plan in the EPMG-E. By early March the process is running at 89% efficiency and within quality specifications.

CONCLUSION

Both companies were making saleable products in March. Unje Enterprises planned for production to start in January and ended up with production in March. Kracon planned for production to start in March and ended up with production in March. Kracon realized early on that production in January could not be reasonably accomplished, whereas Unje saw the difficulty in starting in January and choose to move forward anyway. In the end Unje gained no advantage over Kracon and ended up with more problems and a disorganized construction. Kracon beat Unje to market by 1 week despite taking time to complete upfront design and to work through details.

Moral of the Case Study: This case study shows the advantage that can be gained by taking time to do proper planning and design. Kracon balanced its tripod appropriately, while Unje's tripod is short on time and budget and is severely short in performance.

SOFTWARE PRODUCT DEVELOPMENT PROJECT MANAGEMENT

In order to understand the future of a product, a company has to appreciate the present market as a function of the past. The supply and demand of a product must be choreographed based on what the market currently needs. The defunct Lotus Manuscript Word Processor (ca. 1986) fits the bill as a tool on which present word processing functions were built. This case study is about developing a product to be either one generation ahead (OGA) or multiple generations ahead (MGA) of the marketplace. Lotus Manuscript was a desktop computer tool way ahead of its time. It was the first desktop-based word processor catering the needs of technical documents. In an age when MS Word and Word Perfect were mere document-creating tools, Lotus Manuscript went beyond simplicity. It offered excellent technical word processing capability never before seen at the desktop computer level. The software was so far advanced ahead of its peers that it failed miserably in the market. Its demise after Version 2.0 in 1987 probably fueled the need for its competitors (Word and WordPerfect) to scramble to incorporate technical word processing capabilities.

Lotus Manuscript software was introduced as a tool for the engineering, technical, and scientific community. At a time when MS Word and Word Perfect were still devoid of equation-editing capabilities, several technical manuscripts were actually developed with Lotus Manuscript on the desktop computer. Manuscript's ability to build intricate technical equations appealed a great deal to many in the technical community at that time.

Some of the technically oriented functions offered by Manuscript in 1987 included Screen Capture, Keyboard Stroke Capture; Versatile Common-Language Thesaurus, and Embedded Graphics. Most word processing users of that era did not need those types of capabilities. So, the software mostly languished on the shelves of software stores. The software, however, found ready users among technical professionals—engineers, scientists, and researchers. For the few years that it lasted, Manuscript was the word processor of choice for many people in academia. The technical orientation of the software made it complicated to learn and use. Even trainers cursed the software as they struggled to learn it well enough to teach it to others. To those people, it was a welcome relief when the software took its last breath in the software market around 1989.

The case of Manuscript has important lessons for those engaged in product development. You do not want your product to stray too far ahead of its generation. One generation ahead (OGA) is sufficient as against being multiple generations ahead (MGA). To move too far beyond OGA may spell irrecoverable doom for a product.

Most word processing programs are introduced in functional increments due to hurried development efforts or market pressures. Consequently, we often see quick successions of Versions, 1.0, 1.01, 1.2, … , and so forth. By contrast, Lotus Manuscript was developed as a well-thought-out product in giant steps. Version 1.0 possessed functionality beyond what most users of the era would ever need. Proficient users of Manuscript at that time were generally satisfied with the completeness of the software. Some minor deficiencies that were identified could usually be circumvented by cleverly utilizing the available functions. When Lotus Development Corporation came out with Version 2.0 of Manuscript in 1987, it was not because of the need to fix minor bugs, but rather because of the commitment and desire to incorporate additional giant steps in the software. The relatively few bugs in Version 1.0 make it possible for Lotus to skip the intermediate "bug-fixing" versions that are common in many software packages. This is a good lesson for the present generation of software developers. Unfortunately, market pressures and demands often dictate that a product be rushed to the market before it is really ready for marketing.

For mathematical, scientific, or engineering documents, few low-cost desktop-based word processors of the mid 1980s could rival Lotus Manuscript. For a nontechnical user, the awesome power of Manuscript 2.0 was very intimidating. This might have been a major reason for the market failure of the software. There were (and still are) more nontechnical users of a word processor in the population than there were technical users. The early success of Microsoft Word was because it wooed the more numerous general users. Once the market share was secured, more functionality and versatility were incorporated into the product. Product development professionals should take note of this incremental strategy. This case study is a good guide for developing software faster, better, more efficiently, and more relevant to the current market.

PERRY COMPUTER TECHNOLOGIES

B. B. Perry Technologies is a microcomputer design and manufacturing company that was started in a garage by Bob Perry in 1973. In the early years, Bob ran the company and was largely responsible for making it a huge success. He often worked 20–22 hr a day to accomplish that success. The company grew to a $800 million company in 1980, when Bob sold the company. Substantial stock holdings of the company are still presently held by early company employees, who all have digital electrical engineering backgrounds and who now head the company. Sales of the company peaked at $3 billion in 1984. The company presently employs about 4500 people.

The market has evolved such that almost all competing companies build to a de facto standard (a standard to which Perry products do not adhere), and have gained the majority of the market share. Perry is losing substantial business in both its general-purpose machines and its engineering CAD workstation markets. In attempting to resolve its present problems, two camps within the company have arisen with differing views. The marketing branch of the business is taking the stance that modifying the Perry product to be compatible with the presently popular standards will solve its present problems, while in the design branch it is widely held that the present market for its products can be expanded through constant redoubling of efforts to make the Perry products technologically innovative enough to lead the market. To make matters worse, engineers from the manufacturing environment are now arguing that by cutting manufacturing costs, the present products could be "dumped" on the market at extremely low cost.

After attending a project management seminar, the company heads have decided a project management approach could help in the unification of efforts in future product development, and have set a course of action to internally train personnel in the marketing, design, and manufacturing groups concepts of informal project management.

CASE STUDY DISCUSSIONS

1. What are the current company problems?
2. What can the present project management plan be expected to do?
3. What strategy would you suggest in the company's present circumstances?

Appendix A: Project Terms and Definitions

ABC   Activity-based costing. Bottom-up estimating and summation based on material and labor required for activities making up a project.

Accept   The act of formally receiving or acknowledging a deliverable and regarding it as being true, sound, suitable, or complete.

Acceptance   The act of formally signifying satisfaction with an outcome or a deliverable.

Acceptance Criteria   Those criteria, including performance requirements and essential conditions that must be met before project deliverables are accepted.

Acquire Project Team   The process of obtaining the human resources needed to complete the project.

Activity   A component of work performed during the course of & project. See also schedule activity.

Activity Attributes   Multiple attributes associated with each schedule activity that can be included within the activity list. Activity attributes include activity codes, predecessors, successors, logical relationships, leads and lags, resource requirements, imposed dates, constraints, and assumptions.

Activity-Based Costing   See ABC.

Activity-Based Management   The achievement of strategic objectives and customer satisfaction by managing value added activities.

Activity Code   One or more numerical or text values that identify characteristics of the work or in some way categorize the schedule activity that allows filtering and ordering of activities within reports.

Activity Definition   The process of identifying the specific schedule activities that need to be performed to produce the various project deliverable.

Activity Description   A short phrase or label for each schedule activity used in conjunction with an activity identifier to differentiate that project schedule activity from other schedule activities. The activity description normally describes the scope of work of the schedule activity.

Activity Duration   The time in calendar units between the start and finish of a schedule activity. See also **Actual Duration, Original Duration**, and **Remaining Duration**.

Activity Duration Estimating   The process of estimating the number of work periods that will be needed to complete individual schedule activities.

Activity Identifier   A short unique numeric or text identification assigned to each schedule activity to differentiate that project activity from other activities. Typically unique within any one project schedule network diagram.

Activity List   A documented tabulation of schedule activities that shows the activity description, activity identifier, and a sufficiently detailed scope of work description so project team members understand what work is to be performed.

Activity-on-Arrow (AOA)   A project network diagramming technique in which activities are represented by lines (or arrows) and nodes represent starting and ending points. See **Arrow Diagramming Method, Activity-on-Node (AON), and Precedence Diagramming Method**.

Activity-on-Node (AON)   A project network technique in which nodes represent activities and lines (or arrows) represent precedence relationships.

Activity Resource Estimating   The process of estimating the types and quantities of resources required to perform each schedule activity.

Activity Sequencing   The process of identifying and documenting dependencies among schedule activities.

Actual Cost (AC)   Total costs actually incurred and recorded in accomplishing work performed during a given time period for a schedule activity or work breakdown structure component. Actual cost can sometimes be direct labor hours alone, direct costs alone, or all costs including indirect costs. Also referred to as the actual cost of work performed (ACWP). See also **Earned Value Management and Earned Value Technique**.

Actual Cost of Work Performed (ACWP)   See **Actual Cost (AC)**.

Actual Duration   The time in calendar units between the actual start date of the schedule activity and either the data date of the project schedule if the schedule activity is in progress or the actual finish date if the schedule activity is complete.

Actual Finish   The point in time that work, actually ended on a schedule activity. (Note: In some application areas, the schedule activity is considered "finished" when work is "substantially complete.")

Actual Start   The point in time that work actually started on a schedule activity.

Affinity Diagram   A pictorial clustering of items into similar (or related) categories.

Analogous Estimating   An estimating technique that uses the values of parameters, such as scope, cost, budget, and duration or measures of scale like size, weight, and complexity from a previous, similar activity, as the basis for estimating the same parameter or measure for a future activity. It is frequently used to estimate a parameter when there is a limited amount of detailed information about the project (e.g., in the early phases). Analogous estimating is a form of expert judgment. Analogous estimating is most reliable when the previous activities are similar in fact and not just in appearance, and the project team members preparing the estimates have the needed expertise.

Application Area   A category of projects that have common components significant in such projects but are not needed or present in all projects. Application areas are usually defined in terms of either the product (i.e., by similar technologies or production methods), the type of customer (i.e., internal versus external, government versus commercial), or industry sector (i.e., utilities, automotive, aerospace, information technologies). Application areas can overlap.

Apportioned Effort   Effort applied to project work that is not readily divisible into discrete efforts for that work but is related in direct proportion to measurable discrete work efforts. Contrast with discrete effort.

Approve/Approval   The act of formally confirming, sanctioning, ratifying, or agreeing to something.

Approved Change Request   A change request that has been processed through the integrated change control process and approved. Contrast with requested change.

Arrow   The graphic presentation of a schedule activity in the arrow diagramming method or a logical relationship between schedule activities in the precedence diagramming method.

Arrow Diagramming Method (ADM)   A schedule network diagramming technique in which schedule activities are represented by arrows. The tail of the arrow represents the start, and the head represents the finish of the schedule activity. (The length of the arrow does not represent the expected duration of the schedule activity.) Schedule activities are connected at points called nodes (usually drawn as small circles) to illustrate the sequence in which the schedule activities are expected to be performed. See also **Precedence Diagramming Method.**

As-of Date   See **Data Date**.

Assumptions   Assumptions are factors that, for planning purposes, are considered to be true, real, or certain without proof or demonstration. Assumptions affect all aspects of project planning, and are part of the progressive elaboration of the project. Project teams frequently identify, document, and validate assumptions as part of their planning process. Assumptions generally involve a degree of risk.

Assumptions Analysis   A technique that explores the accuracy of assumptions and identifies risks to the project from inaccuracy, inconsistency, or incompleteness of assumptions.

Authority   The right to apply project resources, expend funds, make decisions, or give approvals.

Backward Pass   The calculation of late finish dates and late start dates for the uncompleted portions of all schedule activities. Determined by working backwards through the schedule network logic from the project's end date. The end date may be calculated in a forward pass or set by the customer or sponsor. See also schedule network analysis.

Bar Chart   A graphic display of schedule-related information. In the typical bar chart, schedule activities or work breakdown structure components are listed down the left side of the chart, dates are shown across the top, and activity durations are shown as date-placed horizontal bars. Also called a Gantt chart.

Baseline   The approved time phased plan (for a project, a work breakdown structure component, a work package, or a schedule activity), plus or minus approved project scope, cost, schedule, and technical changes. Generally refers to the current baseline, but may refer to the original or some other baseline. Usually used with a modifier (e.g., cost baseline, schedule baseline, performance measurement baseline, or technical baseline). See also Performance Measurement Baseline.

Baseline Finish Date   The finish date of a schedule activity in the approved schedule baseline. See also scheduled finish date.

Baseline Start Date   The start date of a schedule activity in the approved schedule baseline. See also scheduled start date.

Best Practices   Processes, procedures, and techniques that have consistently demonstrated achievement of expectations and that are documented for the purposes of sharing, repetition, replication, adaptation, and refinement.

Bill of Materials (BOM)   A documented formal hierarchical tabulation of the physical assemblies, subassemblies, and components needed to fabricate a product.

Bottom-up Estimating   A method of estimating a component of work. The work is decomposed into more detail. An estimate is prepared of what is needed to meet the requirements of each of the lower, more detailed pieces of work, and these estimates are then aggregated into a total quantity for the component of work. The accuracy of bottom-up estimating is driven by the size and complexity of the work identified at the lower levels. Generally smaller work scopes increase the accuracy of the estimates.

Brainstorming   A general data gathering and creativity technique that can be used to identify risks, ideas or solutions to issues by using a group of team members or subject matter experts. Typically, a brainstorming session is structured so that each participant's ideas are recorded for later analysis.

Budget   The approved estimate for the project or any work breakdown structure component or any schedule activity. See also **Estimate**.

Budget at Completion   The sum of all the budget values established for the work to be performed on a project or a work breakdown structure component or a schedule activity. The total planned value for the project.

Budgeted Cost of Work Performed (BCWP)   See **Earned Value (EV)**.

Budgeted Cost of Work Scheduled (BCWS)   See **Planned Value (PV)**.

Buffer   See Reserve.

Buyer   The acquirer of products, services, or results for an organization.

Calendar Unit   The smallest unit of time used in scheduling the project. Calendar units are generally in hours, days, or weeks, but can also be in quarter years, months, shifts, or even in minutes.

Change Control   Identifying, documenting, approving or rejecting, and controlling changes to the project baselines.

Change Control Board   A formally constituted group of stakeholders responsible for reviewing, evaluating, approving, delaying, or rejecting changes to the project, with all decisions and recommendations being recorded.

Change Control System   A collection of formal documented procedures that define how project deliverables and documentation will be controlled, changed, and approved. In most application areas the change control system is a subset of the configuration management system.

Change Request   Requests to expand or reduce the project scope, modify policies, processes, plans, or procedures, modify costs or budgets, or revise schedules. Requests for a change can be direct or indirect, externally or internally initiated, and legally or contractually mandated or optional. Only formally documented requested changes are processed and only approved change requests are implemented.

Chart of Accounts   Any numbering system used to monitor project costs by category (e.g., labor, supplies, materials, and equipment). The project chart of accounts is usually based upon the corporate chart of accounts of the primary performing organization. Contrast with Code of Accounts.

Charter   See **Project Charter**.

Checklist   Items listed together for convenience of comparison, or to ensure the actions associated with them are managed appropriately and not forgotten. An example is a list of items to be inspected that is created during quality planning and applied during quality control.

Claim   A request, demand, or assertion of rights by a seller against a buyer, or vice versa, for consideration, compensation, or payment under the terms of a legally binding contract, such as for a disputed change.

Close Project   The process of finalizing all activities across all of the project process groups to formally close the project or phase.

Closing Processes   Those processes performed to formally terminate all activities of a project or phase, and transfer the completed product to others or close a canceled project.

Code of Accounts   Any numbering system used to uniquely identify each component of the work breakdown structure. Contrast with **Chart of Accounts**.

Colocation   An organizational placement strategy where the project team members are physically located close to one another so as to improve communication, working relationships, and productivity.

Common Cause   A source of variation that is inherent in the system and predictable. On a control chart, it appears as part of the random process variation (i.e., variation from a process that would be considered normal or not unusual), and is indicated by a random pattern of points within the control limits. Also referred to as random cause. Contrast with special cause.

Communication   A process through which information is exchanged among persons using a common system of symbols, signs, or behaviors.

Communication Management Plan   The document that describes the communications needs and expectations for the project, how and in what format information will be communicated, when and where each communication will be made, and who is responsible for providing each type of communication. A communication management plan can be formal or informal, highly detailed or broadly framed, based on the requirements of the project stakeholders. The communication management plan is contained in, or is a subsidiary plan of, the project management plan.

Communications Planning   The process of determining the information and communications needs of the project stakeholders: who they are, their levels of interest and influence on the project, who needs what information, when will they need it, and how it will be given to them.

Compensation   Something given or received, a payment or recompense, usually something monetary or in kind for products, services, or results provided or received.

Component   A constituent part, element, or piece of a complex whole.

Configuration Management System   A subsystem of the overall project management system. It is a collection of formal documented procedures used to apply technical and administrative direction and surveillance to identify and document the functional and physical characteristics of a product, result, service, or component; control any changes to such characteristics; record and report each change and its implementation status; and support the audit of the products, results, or components to verify conformance to requirements. It includes the documentation, tracking systems, and defined approval levels necessary for authorizing and controlling changes. In most application areas, the configuration management system includes the change control system.

Constraint The state, quality, or sense of being restricted to a given course of action or inaction. An applicable restriction or limitation, either internal or external to the project, that will affect the performance of the project or a process. For example, a schedule constraint is any limitation or restraint placed on the project schedule that affects when a schedule activity can be scheduled and is usually in the form of fixed imposed dates. A cost constraint is any limitation or restraint placed on the project budget such as funds available over time. A project resource constraint is any limitation or restraint placed on resource usage, such as what resource skills or disciplines are available and the amount of a given resource available during a specified time frame.

Contingency See Reserve.

Contingency Allowance See Reserve.

Contingency Reserve The amount of funds, budget, or time needed above the estimate to reduce the risk of overruns of project objectives to a level acceptable to the organization.

Contract A contract is a mutually binding agreement that obligates the seller to provide the specified product or service or result and obligates the buyer to pay for it.

Contract Administration The process of managing the contract and the relationship between the buyer and seller, reviewing and documenting how a seller is performing or has performed to establish required corrective actions and provide a basis for future relationships with the seller, managing contract-related changes, and, when appropriate, managing the contractual relationship with the outside buyer of the project.

Contract Closure The process of completing and settling the contract, including resolution of any open items and closing each contract.

Contract Management Plan The document that describes how a specific contract will be administered and can include items such as required documentation delivery and performance requirements. A contract management plan can be formal or informal, highly detailed or broadly framed, based on the requirements in the contract. Each contract management plan is a subsidiary plan of the project management plan.

Contract SOW A narrative description of products, services, or results to be supplied under contract.

Contract Work Breakdown Structure (CWBS) A portion of the work breakdown structure for the project developed and maintained by a seller contracting to provide a subproject or project component.

Control Comparing actual performance with planned performance, analyzing variances, assessing trends to effect process improvements, evaluating possible alternatives, and recommending appropriate corrective action as needed.

Control Account (CA) A management control point where the integration of scope, budget, actual cost, and schedule takes place, and where the measurement of performance will occur. Control accounts are placed at selected management points (specific components at selected levels) of the work breakdown structure. Each control account may include one or more work packages, but each work package may be associated with only one control account. Each control account is associated with a specific single organizational component in the organizational breakdown structure (OBS). Previously called a cost account. See also Work Package.

Control Account Plan (CAP) A plan for all the work and effort to be performed in a control account. Each CAP has a definitive statement of work, schedule, and time-phased budget. Previously called a cost account plan.

Control Chart A graphic display of process data over time and against established control limits, and has a centerline that assists in detecting a trend of plotted values toward either control limit.

Control Limits The area composed of three standard deviations on either side of the centerline, or mean, of a normal distribution of data plotted on a control chart that reflects the expected variation in the data. See also **Specification Limits**.

Controlling See **Control**.

Corrective Action Documented direction for executing the project work to bring expected future performance of the project work in line with the project management plan.

Cost The monetary value or price of a project activity or component that includes the monetary worth of the resources required to perform and complete the activity or component, or to produce the component. A specific cost can be composed of a combination of cost components including direct labor hours, other direct costs, indirect labor hours, other indirect costs, and purchased price. However, in the earned value management methodology, in some instances, the term cost can represent only labor hours without conversion to monetary worth. See also **Actual Cost** and **Estimate**.

Cost Baseline See **Baseline**.

Cost Budgeting The process of aggregating the estimated costs of individual activities or work packages to establish a cost baseline.

Cost Control The process of influencing the factors that create variances and controlling changes to the project budget.

Cost Estimating The process of developing an approximation of the cost of the resources needed to complete project activities.

Cost Management Plan The document that sets out the format and establishes the activities and criteria for planning, structuring, and controlling the project costs. A cost management plan can be formal or informal, highly detailed or broadly framed, based on the requirements of the project stakeholders. The cost management plan is contained in, or is a subsidiary plan of, the project management plan.

Cost of Quality (COQ) Determining the costs incurred to ensure quality. Prevention and appraisal costs (cost of conformance) include costs for quality planning, quality control (QC), and quality assurance to ensure compliance to requirements (i.e., training, QC systems, etc.). Failure costs (cost of nonconformance) include costs to rework products, components, or processes that are noncompliant, costs of warranty work and waste, and loss of reputation.

Cost Performance Index (CPI)   A measure of cost efficiency on a project. It is the ratio of earned value (EV) to actual costs (AC), that is, CPI = EV/AC. A CPI value equal to or greater than one indicates a favorable condition and a value less than one indicates an unfavorable condition.

Cost-Plus-Fee (CPF)   A type of cost reimbursable contract where the buyer reimburses the seller for seller's allowable costs for performing the contract work and seller also receives a fee calculated as an agreed upon percentage of the costs. The fee varies with the actual cost.

Cost-Plus-Fixed-Fee (CPFF) Contract   A type of cost reimbursable contract in which the buyer reimburses the seller for the seller's allowable costs (allowable costs are defined by the contract) plus a fixed amount of profit (fee).

Cost-Plus-Incentive-Fee (CPIF) Contract   A type of cost-reimbursable contract where the buyer reimburses the seller for the seller's allowable costs (allowable costs are defined by the contract), and the seller earns its profit if it meets defined performance criteria.

Cost-Plus-Percentage of Cost (CPPC)   See **Cost-Plus-Fee**.

Cost Reimbursable Contract   A type of contract involving payment (reimbursement) by the buyer to the seller for the seller's actual costs, plus a fee typically representing seller's profit. Costs are usually classified as direct or indirect costs. Direct costs are those incurred for the exclusive benefit of the project, such as salaries of full-time project staff. Indirect costs, also called overhead and general and administrative cost, are those allocated to the project by the performing organization as a cost of doing business, such as salaries of management indirectly involved in the project, and cost of electric utilities for the office. Indirect costs are usually calculated as a percentage of direct costs. Cost reimbursable contracts often include incentive clauses where, if the seller meets or exceeds selected project objectives, such as schedule targets or total cost, then the seller receives from the buyer an incentive or bonus payment.

Cost Variance (CV)   A measure of cost performance on a project. It is the algebraic difference between earned value (EV) and actual cost (AC), that is, CV = EV − AC. A positive value indicates a favorable condition and a negative value indicates an unfavorable condition.

Crashing   A specific type of project schedule compression technique performed by taking action to decrease the total project schedule duration after analyzing a number of alternatives to determine how to get the maximum schedule duration compression for the least additional cost. Typical approaches for crashing a schedule include reducing schedule activity durations and increasing the assignment of resources on schedule activities. See Schedule Compression and Fast Tracking.

Create Work Breakdown Structure   The process of subdividing the major project deliverables and project work into smaller, more manageable components.

Criteria   Standards, rules, or tests on which a judgment or decision can be based, or by which a product, service, result, or process can be evaluated.

Critical Activity   Any schedule activity on a critical path in a project schedule. Most commonly determined by using the Critical Path Method. Although some activities are "critical," in the dictionary sense, without being on the critical path, this meaning is seldom used in the project context.

Critical Chain Method   A schedule network analysis technique that modifies the project schedule to account for limited resources. The critical chain method mixes deterministic and probabilistic approaches to schedule network analysis.

Critical Path   Generally, but not always, the sequence of schedule activities that determines the duration of the project. Generally, it is the longest path through the project. However, a critical path can end, as an example, on a schedule milestone that is in the middle of the project schedule and that has a finish-no-later-than imposed date schedule constraint. See also **Critical Path Method**.

Critical Path Method (CPM)   A schedule network analysis technique used to determine the amount of scheduling flexibility (the amount of float) on various logical network paths in the project schedule network, and to determine the minimum total project duration. Early start and finish dates are calculated by means of forward pass, using a specified start date. Late start and finish dates are calculated by means of a backward pass, starting from a specified completion date, which sometimes is the project early finish date determined during the forward-pass calculation.

Current Finish Date   The current estimate of the point in time when a schedule activity will be completed, where the estimate reflects any reported work progress. See also **Scheduled Finish Date** and **Baseline Finish Date**.

Current Start Date   The current estimate of the point in time a schedule activity will begin, where the estimate reflects any reported work progress. See also **Scheduled Start Date** and **Baseline Start Date**.

Customer   The person or organization that will use the project's product or service or result. See also User.

Data Date (DD)   The date up to or through which the project's reporting system has provided actual status and accomplishments. In some reporting systems, the status information for the data date is included in the past and in some systems the status information is in the future. Also called as-of-date and time-now date.

Date   A term representing the day, month, and year of a calendar, and, in some instances, the time of day.

Decision Tree Analysis   The decision tree is a diagram that describes a decision under consideration and the implications of choosing one or another of the available alternatives. It is used when some future scenarios or outcomes of actions are uncertain. It incorporates probabilities and the costs or rewards of each logical path of events and future decisions, and uses expected monetary value analysis to help the organization identify the relative values of alternate actions. See also **Expected Monetary Value Analysis**.

Decompose   See **Decomposition**.

Decomposition   A planning technique that subdivides the project scope and project deliverables into smaller, more manageable components until the project work associated with accomplishing the project scope and providing the deliverables is defined in sufficient detail to support executing, monitoring, and controlling the work.

Defect   An imperfection or deficiency in a project component where that component does not meet its requirements or specifications and needs to be either repaired or replaced.

Defect Repair   Formally documented identification of a defect in a project component with a recommendation to either repair the defect or completely replace the component.

Deliverable   Any unique and verifiable product, result, or capability to perform a service that must be produced to complete a process, phase, or project. Often used more narrowly in reference to an external deliverable, which is a deliverable that is subject to approval by the project sponsor or customer. See also **Product, Service, and Result**.

Delphi Technique   An information-gathering technique used as a way to reach a consensus of experts on a subject. Experts on the subject participate in this technique anonymously. A facilitator uses a questionnaire to solicit ideas about the important project points related to the subject. The responses are summarized and are then recirculated to the experts for further comment. Consensus may be reached in a few rounds of this process. The Delphi technique helps reduce bias in the data and keeps any one person from having undue influence on the outcome.

Dependency   See **Logical Relationship**.

Design Review   A management technique used for evaluating a proposed design to ensure that the design of the system or product meets the customer requirements or to assure that the design will perform successfully, can be produced, and can be maintained.

Develop Project Charter   The process of developing the project charter that formally authorizes a project.

Develop Project Management Plan   The process of documenting the actions necessary to define, prepare, integrate, and coordinate all subsidiary plans into a project management plan.

Develop Project Scope Statement (Preliminary)   The process of developing the preliminary project scope statement that provides a high-level scope narrative.

Develop Project Team   The process of improving the competencies and interaction of team members to enhance project performance.

Direct and Manage Project Execution   The process of executing the work, defined in the project management plan to achieve the project's requirements defined in the project scope statement.

Discipline   A field of work requiring specific knowledge and that has a set of rules governing work conduct (e.g., mechanical engineering, computer programming, cost estimating, etc.).

Discrete Effort   Work effort that is directly identifiable to the completion of specific work breakdown structure components and deliverables, and that can be directly planned and measured. Contrast with apportioned effort.

Document   A medium, and the information recorded thereon, that generally has permanence and can be read by a person or a machine. Examples include project management plans, specifications, procedures, studies, and manuals.

Documented Procedure   A formalized written description of how to carry out an activity, process, technique, or methodology.

Dummy Activity A schedule activity of zero duration used to show a logical relationship in the arrow diagramming method. Dummy activities are used when logical relationships cannot be completely or correctly described with schedule activity arrows. Dummy activities are generally shown graphically as a dashed line headed by an arrow.

Duration The total number of work periods (not including holidays or other non-working periods) required to complete a schedule activity or work breakdown structure component. Usually expressed as workdays or workweeks. Sometimes incorrectly equated with elapsed time. Contrast with effort. See also **Original Duration, Remaining Duration, and Actual Duration**.

Early Finish Date (EF) In the critical path method, the earliest possible point in time on which the uncompleted portions of a schedule activity (or the project) can finish, based on the schedule network, logic, the data date, and any schedule constraints. Early finish dates can change as the project progresses and as changes are made to the project management plan.

Early Start Date (ES) In the critical path method, the earliest possible point in time the uncompleted portions of a schedule activity (or the project) can start, based on the schedule network logic, the data date, and any schedule constraints. Early start dates can change as the project progresses and as changes are made to the project management plan.

Earned Value (EV) The value of completed work expressed in terms of the approved budget assigned to that work for a schedule activity or work breakdown structure component. Also referred to as the budgeted cost of work performed (BCWP).

Earned Value Management (EVM) A management methodology for integrating scope, schedule, and resources, and for objectively measuring project performance and progress. Performance is measured by determining the budgeted cost of work performed (i.e., earned value) and comparing it to the actual cost of work performed (i.e., actual cost). Progress is measured by comparing the earned value to the planned value.

Earned Value Technique (EVT) A specific technique for measuring the performance of work for a work breakdown structure component, control account, or project. Also referred to as the earning rules and crediting method.

Effort The number of labor units required to complete a schedule activity or work breakdown structure component. Usually expressed as staff hours, staff days, or staff weeks. Contrast with duration.

Enterprise A company, business, firm, partnership, corporation, or governmental agency.

Enterprise Environmental Factors Any or all external environmental factors and internal organizational environmental factors that surround or influence the project's success. These factors are from any or all of the enterprises involved in the project, and include organizational culture and structure, infrastructure, existing resources, commercial databases, market conditions, and project management software.

Estimate A quantitative assessment of the likely amount or outcome. Usually applied to project costs, resources, effort, and durations and is usually preceded by a modifier (i.e., preliminary, conceptual, feasibility, order-of-magnitude, and definitive). It should always include some indication of accuracy (e.g., plus or minus percent).

Estimate at Completion (EAC)   The expected total cost of a schedule activity, a work breakdown structure component, or the project when the defined scope of work will be completed. EAC is equal to the actual cost (AC) plus the estimate to complete (ETC) for all of the remaining work. EAC = AC + ETC. The EAC may be calculated based on performance to date or estimated by the project team based on other factors, in which case it is often referred to as the latest revised estimate. See also **Earned Value Technique** and **Estimate to Complete**.

Estimate to Complete (ETC)   The expected cost needed to complete all the remaining work for a schedule activity, work breakdown structure component, or the project. See also **Earned Value Technique** and **Estimate at Completion**.

Event   Something that happens—an occurrence, an outcome.

Exception Report   Document that includes only major variations from the plan (rather than all variations).

Execute   Directing, managing, performing, and accomplishing the project work, providing the deliverables, and providing work performance information.

Executing   See Execute.

Executing Processes   Those processes performed to complete the work defined in the project management plan to accomplish the project's objectives defined in the project scope statement.

Execution   See Execute.

Expected Monetary Value (EMV) Analysis   A statistical technique that calculates the average outcome when the future includes scenarios that may or may not happen. A common use of this technique is within decision tree analysis. Modeling and simulation are recommended for cost and schedule risk analysis because it is more powerful and less subject to misapplication than expected monetary value analysis.

Expert Judgment   Judgment provided based on expertise in an application area, knowledge area, discipline, industry, and so forth. as appropriate for the activity being performed. Such expertise may be provided by any group or person with specialized education, knowledge, skill, experience, or training, and is available from many sources, including other units within the performing organization; consultants; stakeholders, including customers, professional and technical associations; and industry groups.

Failure Mode and Effect Analysis (FMEA)   An analytical procedure in which each potential failure mode in every component of a product is analyzed to determine its effect on the reliability of that component and, by itself or in combination with other possible failure modes, on the reliability of the product or system and on the required function of the component; or the examination of a product (at the system and/or lower levels) for all ways that a failure may occur. For each potential failure, an estimate is made of its effect on the total system and of its impact. In addition, a review is undertaken of the action planned to minimize the probability of failure and to minimize its effects.

Fast Tracking   A specific project schedule compression technique that changes network logic to overlap phases that would normally be done in sequence, such as the design phase and construction phase, or to perfor m schedule activities in parallel. See **Schedule Compression** and **Crashing**.

Finish Date   A point in time associated with a schedule activity's completion. Usually qualified by one of the following: actual, planned, estimated, scheduled, early, late, baseline, target, or current.

Finish-to-Finish (FF)   The logical relationship where completion of work of the successor activity cannot finish until the completion of work of the predecessor activity. See also **Logical Relationship**.

Finish-to-Start (FS)   The logical relationship in which initiation of work of the successor activity depends upon the completion of work of the predecessor activity. See also Logical Relationship.

Firm-Fixed-Price (FFP) Contract   A type of fixed-price contract where the buyer pays the seller a set amount (as defined by the contract), regardless of the seller's costs.

Fixed-Price-Incentive-Fee (FPIF) Contract   A type of contract where the buyer pays the seller a set amount (as defined by the contract), and the seller can earn an additional amount if the seller meets defined performance criteria.

Fixed-Price or Lump-Sum Contract   A type of contract involving a fixed total price for a well-defined product. Fixed-price contracts may also include incentives for meeting or exceeding selected project objectives, such as schedule targets. The simplest form of a fixed price contract is a purchase order.

Float   Also called slack. See **Total Float** and **Free Float**.

Flowcharting   The depiction in a diagram format of the inputs, process actions, and outputs of one or more processes within a system.

Forecasts   Estimates or predictions of conditions and events in the project's future based on information and knowledge available at the time of the forecast. Forecasts are updated and reissued based on work performance information provided as the project is executed. The information is based on the project's past performance and expected future performance, and includes information that could impact the project in the future, such as estimate at completion and estimate to complete.

Forward Pass   The calculation of the early start and early finish dates for the uncompleted portions of all network activities. See also **Schedule Network Analysis** and **Backward Pass**.

Free Float (FF)   The amount of time that a schedule activity can be delayed without delaying the early start of any immediately following schedule activities. See also **Total Float**.

Functional Manager   Someone with management authority over an organizational unit within a functional organization. The manager of any group that actually makes a product or performs a service. Sometimes called a line manager.

Functional Organization   A hierarchical organization where each employee has one clear superior, staff are grouped by areas of specialization, and managed by a person with expertise in that area.

Funds   A supply of money or pecuniary resources immediately available.

Gantt Chart   See **Bar Chart**.

Goods   Commodities, wares, or merchandise.

Grade   A category or rank used to distinguish items that have the same functional use but do not share the same requirements for quality.

Ground Rules   A list of acceptable and unacceptable behaviors adopted by a project team to improve working relationships, effectiveness, and communication.

Hammock Activity   See **Summary Activity**.

Historical Information   Documents and data on prior projects including project files, records, correspondence, closed contracts, and closed projects.

Human Resource Planning   The *process* of identifying and documenting *project roles,* responsibilities and reporting relationships, as well as creating the *staffing management plan.*

Imposed Date   A fixed date imposed on a schedule activity or schedule milestone, usually in the form of a "start no earlier than" and "finish no later than" date.

Influence Diagram   Graphical representation of situations showing causal influences, time ordering of events, and other relationships among variables and outcomes.

Influencer   Persons or groups that are not directly related to the acquisition or use of the projects and products but, due to their position in the customer organization, can influence, positively or negatively, the course of the project.

Information Distribution   The process of making needed information available to project stakeholders in a timely manner.

Initiating Processes   Those processes performed to authorize and define the scope of a new phase or project or that can result in the continuation of halted project work. A large number of the initiating processes are typically done outside the project's scope of control by the organization, program, or portfolio processes and those processes provide input to the project's initiating processes group.

Initiator   A person or organization that has both the ability and authority to start a project.

Input   Any item, whether internal or external to the project, that is required by a process before that process proceeds. An input may be an output from a predecessor process.

Inspection   Examining or measuring to verify whether an activity, component, product, result, or service conforms to specified requirements.

Integral   Essential to completeness; requisite; constituent with; formed as a unit with another component.

Integrated   Interrelated, interconnected, interlocked, or enmeshed components blended and unified into a functioning or unified whole.

Integrated Change Control   The process of reviewing all change requests, approving changes and controlling changes to deliverables and organizational process assets.

Invitation for Bid (IFB)   Generally, this term is equivalent to request for proposal. However, in some application areas, it may have a narrower or more specific meaning.

Issue   A point or matter in question or in dispute, or a point or matter that is not settled and is under discussion or over which there are opposing views or disagreements.

Knowledge Knowing something with the familiarity gained through experience, education, observation, or investigation.

Knowledge Area Process An identifiable project management process within a knowledge area.

Knowledge Area, Project Management See Project Management Knowledge Area.

Lag A modification of a logical relationship that directs a delay in the successor activity. For example, in a finish-to-start dependency with a 10-day lag, the successor activity cannot start until ten days after the predecessor activity has finished. See also **Lead**.

Late Finish Date (LF) In the Critical Path Method, the latest possible point in time that a schedule activity may be completed based on the schedule network logic, the project completion date, and any constraints assigned to the schedule activities without violating a schedule constraint or delaying the project completion date. The late finish dates are determined during the backward pass calculation of the project schedule network.

Late Start Date (LS) In the critical path method, the latest possible point in time that a schedule activity may begin based on the schedule network logic, the project completion date, and any constraints assigned to the schedule activities without violating a schedule constraint or delaying the project completion date. The late start dates are determined during the backward pass calculation of the project schedule network.

Latest Revised Estimate See **Estimate at Completion**.

Lead A modification of a logical relationship that allows an acceleration of the successor activity. For example, in a finish-to-start dependency with a ten-day lead, the successor activity can start ten days before the predecessor activity has finished. See also **Lag**. A negative lead is equivalent to a positive lag.

Lessons Learned The learning gained from the process of performing the project. Lessons learned may be identified at any point. Also considered a project record, to be included in the lessons learned knowledge base.

Lessons Learned Knowledge Base A store of historical information and lessons learned about both the outcomes of previous project selection decisions and previous project performance.

Level of Effort (LOE) Support-type activity (e.g., seller or customer liaison, project cost accounting, project management, etc.) that does not readily lend itself to measurement of discrete accomplishment. It is generally characterized by a uniform rate of work, performance over a period of time determined by the activities supported.

Leveling See **Resource Leveling**.

Life Cycle See **Project Life Cycle**.

Log A document used to record and describe or denote selected items identified during execution of a process or activity. Usually used with a modifier, such as issue, quality control, action, or defect.

Logic See **Network Logic**.

Logic Diagram See **Project Schedule Network Diagram**.

Logical Relationship A dependency between two project schedule activities, or between a project schedule activity and a schedule milestone. See also

Precedence Relationship. The four possible types of logical relationships are finish-to-start, finish-to-finish, start-to-start, and start-to-finish.

Manage Project Team   The process of tracking team member performance, providing feedback, resolving issues, and coordinating changes to enhance project performance.

Manage Stakeholders   The process of managing communications to satisfy the requirements of, and resolve issues with, project stakeholders.

Master Schedule   A summary-level project schedule that identifies the major deliverable and work breakdown structure components and key schedule milestones. See also **Milestone Schedule**.

Materiel   The aggregate of things used by an organization in any undertaking, such as equipment, apparatus, tools, machinery, gear, material, and supplies.

Matrix Organization   Any organizational structure in which the project manager shares responsibility with the functional managers for assigning priorities and for directing the work of persons assigned to the project.

Methodology   A system of practices, techniques, procedures, and rules used by those who work in a discipline.

Milestone   A significant point or event in the project. See also **Schedule Milestone**.

Milestone Schedule   A summary-level schedule that identifies the major schedule milestones. See also **Master Schedule**.

Monitor   Collect project performance data with respect to a plan, produce performance measures, and report and disseminate performance information.

Monitor and Control Project Work   The process of monitoring and controlling the processes required to initiate, plan, execute, and close a project to meet the performance objectives defined in the project management plan and project scope statement.

Monitoring   See **Monitor**.

Monitoring and Controlling Processes   Those processes performed to measure and monitor project execution so that corrective action can be taken when necessary to control the execution of the phase or project.

Monte Carlo Analysis   A technique that computes, or iterates, the project cost or project schedule many times using input values selected at random from probability distributions of possible costs or durations to calculate a distribution of possible total project cost or completion dates.

Near-Critical Activity   A schedule activity that has low total float. The concept of near-critical is equally applicable to a schedule activity or schedule network path. The limit below which total float is considered near critical is subject to expert judgment and varies from project to project.

Network   See **Project Schedule Network Diagram**.

Network Analysis   See **Schedule Network Analysis**.

Network Logic   The collection of schedule activity dependencies that makes up a project schedule network diagram.

Network Loop   A schedule network path that passes the same node twice. Network loops cannot be analyzed using traditional schedule network analysis techniques such as critical path method.

Network Open End   A schedule activity without any predecessor activities or successor activities creating an unintended break in a schedule network path. Network open ends are usually caused by missing logical relationships.

Network Path   Any continuous series of schedule activities connected with logical relationships in a project schedule network diagram.

Networking   Developing relationships with persons who may be able to assist in the achievement of objectives and responsibilities.

Node   One of the defining points of a schedule network; a junction point joined to some or all of the other dependency lines. See also **Arrow Diagramming Method** and **Precedence Diagramming Method**.

Objective   Something toward which work is to be directed, a strategic position to be attained, or a purpose to be achieved, a result to be obtained, a product to be produced, or a service to be performed.

Operations   An organizational function performing the ongoing execution of activities that produce the same product or provide a repetitive service. Examples are production operations, manufacturing operations, and accounting operations.

Opportunity   A condition or situation favorable to the project, a positive set of circumstances, a positive set of events, a risk that will have a positive impact on project objectives, or a possibility for positive changes. Contrast with **Threat**.

Organization   Group of persons organized for some purpose or to perform some type of work within an enterprise.

Organization Chart   A method for depicting interrelationships among a group of persons working together toward a common objective.

Organizational Breakdown Structure (OBS)   A hierarchically organized depiction of the project organization arranged so as to relate the work packages to the performing organizational units. (Sometimes OBS is expanded as organization breakdown structure with the same definition.)

Organizational Process Assets   Any or all process related assets, from any or all of the organizations involved in the project that are or can be used to influence the project's success. These process assets include formal and informal plans, policies, procedures, and guidelines. The process assets also include the organization's knowledge bases such as lessons learned and historical information.

Original Duration (OD)   The activity duration originally assigned to a schedule activity and not updated as progress is reported on the activity. Typically used for comparison with actual duration and remaining duration when reporting schedule progress.

Output   A product, result, or service generated by s process. May be an input to a successor process.

Parametric Estimating   An estimating technique that uses a statistical relationship between historical data and other variables (e.g., square footage in construction, lines of code in software development) to calculate an estimate for activity parameters, such as scope, cost, budget, and duration. This technique can produce higher levels of accuracy depending upon the sophistication and the underlying data built into the model. An example for the cost parameter is multiplying the planned quantity of work to be performed by the historical cost per unit to obtain the estimated cost.

Pareto Chart   A histogram, ordered by frequency of occurrence, that shows how many results were generated by each identified cause.

Path Convergence   The merging or joining of parallel schedule network paths into the same node in a project schedule network diagram. Path convergence is characterized by a schedule activity with more than one predecessor activity.

Path Divergence   Extending or generating parallel schedule network paths from the same node in a project schedule network diagram. Path divergence is characterized by a schedule activity with more than one successor activity.

Percent Complete (PC or PCT)   An estimate, expressed as a percent, of the amount of work that has been completed on an activity or a work breakdown structure component.

Perform Quality Assurance (QA)   The process of applying the planned, systematic quality activities (such as audits or peer reviews) to ensure that the project employs all processes needed to meet requirements.

Perform Quality Control (QC)   The process of monitoring specific project results to determine whether they comply with relevant quality standards and identifying ways to eliminate causes of unsatisfactory performance.

Performance Measurement Baseline   An approved plan for the project work against which project execution is compared and deviations are measured for management control. The performance measurement baseline typically integrates scope, schedule, and cost parameters of a project, but may also include technical and quality parameters.

Performance Reporting   The process of collecting and distributing performance information. This includes status reporting, progress measurement, and forecasting.

Performance Reports   Documents and presentations that provide organized and summarized work performance information, earned value management parameters and calculations, and analyses of project work progress and status. Common formats for performance reports include bar charts, S-curves, histograms, tables, and project schedule network diagram showing current schedule status.

Performing Organization   The enterprise whose personnel are most directly involved in doing the work of the project.

Phase   See **Project Phase**.

Plan Contracting   The process of documenting the products, services, and results requirements and identifying potential sellers.

Plan Purchases and Acquisitions   The process of determining what to purchase or acquire, and determining when and how to do so.

Planned Finish Date   See **Schedule Finish Date**.

Planned Start Date   See **Scheduled Start Date**.

Planned Value   The authorized budget assigned to the scheduled work to be accomplished for a schedule activity or work breakdown structure component. Also referred to as the budgeted cost of work scheduled (BCWS).

Planning Package   A WBS component below the control account with known work content but without detailed schedule activities. See also **Control Account**.

Planning Processes Those processes performed to define and mature the project scope, develop the project management plan, and identify and schedule the project activities that occur within the project.

Portfolio A collection of projects or programs and other work that are grouped together to facilitate effective management of that work to meet strategic business objectives. The projects or programs of the portfolio may not necessarily be interdependent or directly related.

Portfolio Management The centralized management of one or more portfolios, which includes identifying, prioritizing, authorizing, managing, and controlling projects, programs, and other related work, to achieve specific strategic business objectives.

Position Description An explanation of a project team member's roles and responsibilities.

Practice A specific type of professional or management activity that contributes to the execution of a process and that may employ one or more techniques and tools.

Precedence Diagramming Method (PDM) A schedule network diagramming technique in which schedule activities are represented by boxes (or nodes). Schedule activities are graphically linked by one or more logical relationships to show the sequence in which the activities are to be performed.

Precedence Relationship The term used in the precedence diagramming method for a logical relationship. In current usage, however, precedence relationship, logical relationship, and dependency are widely used interchangeably, regardless of the diagramming method used.

Predecessor Activity The schedule activity that determines when the logical successor activity can begin or end.

Preventive Action Documented direction to perform an activity that can reduce the probability of negative consequences associated with project risks.

Probability and Impact Matrix A common way to determine whether a risk is considered low, moderate, or high by combining the two dimensions of a risk: its probability of occurrence and its impact on objectives if it occurs.

Procedure A series of steps followed in a regular definitive order to accomplish something.

Process A set of interrelated actions and activities performed to achieve a specified set of products, results, or services.

Process Group See **Project Management Process Groups**.

Procurement Documents Those documents utilized in bid and proposal activities, which include buyer's invitation for bid, invitation for negotiations, request for information, request for quotation, request for proposal, and seller's responses.

Procurement Management Plan The document that describes how procurement processes from developing procurement documentation through contract closure will be managed.

Product An artifact that is produced, is quantifiable, and can be either an end item in itself or a component item. Additional words for products are materiel and goods. Contrast with result and service. See also Deliverable.

Product Life Cycle A collection of generally sequential, nonoverlapping product phases whose name and number are determined by the manufacturing and control needs of the organization. The last product life-cycle phase for a product is generally the product's deterioration and death. Generally, a project life cycle is contained within one or more product life cycles.

Product Scope   The features and functions that characterize a product, service, or result.

Product Scope Description   The documented narrative description of the product scope.

Program   A group of related projects managed in a coordinated way to obtain benefits and control not available from managing them individually. Programs may include elements of related work outside of the scope of the discrete projects in the program.

Program Management   The centralized coordinated management of a program to achieve the program's strategic objectives and benefits.

Program Management Office (PMO)   The centralized management of a particular program or programs such that corporate benefit is realized by the sharing of resources, methodologies, tools, and techniques, and related high-level project management focus. See also project management office.

Progressive Elaboration   Continuously improving and detailing a plan as more detailed and specific information and more accurate estimates become available as the project progresses, and thereby producing more accurate and complete plans that result from the successive iterations of the planning process.

Project   A temporary endeavor undertaken to create a unique product, service, or result.

Project Calendar   A calendar of working days or shifts that establishes those dates on which schedule activities are worked and nonworking days that determine those dates on which schedule activities are idle. Typically defines holidays, weekends, and shift hours. See also **Resource Calendar**.

Project Charter   A document issued by the project initiator or sponsor that formally authorizes the existence of a project, and provides the project manager with the authority to apply organizational resources to project activities.

Project Initiation   Launching a process that can result in the authorization and scope definition of a new project.

Project Life Cycle   A collection of generally sequential project phases whose name and number are determined by the control needs of the organization or organizations involved in the project. A life cycle can be documented with a methodology.

Project Management (PM)   The application of knowledge, skills, tools, and techniques to project activities to meet the project requirements.

Project Management Body of Knowledge (PMBOK®)   An inclusive term that describes the sum of knowledge within the profession of project management. As with other professions such as law, medicine, and accounting, the body of knowledge rests with the practitioners and academics that apply and advance it. The complete project management body of knowledge includes proven traditional practices that are widely applied and innovative practices that are emerging in the project management profession.

Project Management Information System (PMIS)   An information system consisting of the tools and techniques used to gather, integrate, and disseminate the outputs of project management processes. It is used to support all aspects of the project from initiating through closing, and can include both manual and automated systems.

Project Management Knowledge Area   An identified area of project management defined by its knowledge requirements and described in terms of its component processes, practices, inputs, outputs, tools, and techniques.

Project Management Office (PMO)   An organizational body or entity assigned various responsibilities related to the centralized and coordinated management of those projects under its domain. The responsibilities of a PMO can range from providing project management support functions to actually being responsible for the direct management of a project. See also program management office.

Project Management Plan   A formal, approved document that defines how the projected is executed, monitored, and controlled. It may be summary or detailed and may be composed of one or more subsidiary management plans and other planning documents.

Project Management Process   One of the processes unique to project management and described in the PMBOK® Guide.

Project Management Process Group   A logical grouping of the project management processes described in the PMBOK® Guide. The project management process groups include initiating processes, planning processes, executing processes, monitoring and controlling processes, and closing processes. Collectively, these five groups are required for any project, have clear internal dependencies, and must be performed in the same sequence on each project, independent of the application area or the specifics of the applied project life cycle. Project management process groups are not project phases.

Project Management Professional (PMP®)   A person certified as a PMP® by the Project Management Institute (PMI®).

Project Management Software   A class of computer software applications specifically designed to aid the project management team with planning, monitoring, and controlling the project, including cost estimating, scheduling, communications, collaboration, configuration management, document control, records management, and risk analysis.

Project Management System   The aggregation of the processes, tools, techniques, methodologies, resources, and procedures to manage a project. The system is documented in the project management plan and its content will vary depending on the application area, organizational influence, complexity of the project, and the availability of existing systems. A project management system, formal or informal, aids a project manager in effectively guiding a project to completion. A project management system is a set of processes and the related monitoring and control functions that are consolidated and combined into a functioning, unified whole.

Project Management Team   The members of the project team who are directly involved in project management activities. On some smaller projects, the project management team may include virtually all of the project team members.

Project Manager (PM)   The person assigned by the performing organization to achieve the project objectives.

Project Organization Chart   A document that graphically depicts the project team members and their interrelationships for a specific project.

Project Phase   A collection of logically related project activities, usually culminating in the completion of a major deliverable. Project phases (also called phases) are mainly completed sequentially, but can overlap in some project situations. Phases can be subdivided into subphases and then components; this hierarchy, if the project or portions of the project are divided into phases, is contained in the work breakdown structure. A project phase is a component of a project life cycle. A project phase is not a project management process group.

Project Process Groups   The five process groups required for any project that have clear dependencies and that are required to be performed in the same sequence on each project, independent of the application area or the specifics of the applied project life cycle. The process groups are initiating, planning, executing, monitoring and controlling, and closing.

Project Schedule   The planned dates for performing schedule activities and the planned dates for meeting schedule milestones.

Project Schedule Network Diagram   Any schematic display of the logical relationships among the project schedule activities. Always drawn from left to right to reflect project work chronology.

Project Scope   The work that must be performed to deliver and product, service, or result with the specified features and functions.

Project Scope Management Plan   The document that describes how the project scope will be defined, developed, and verified and how the work breakdown structure will be created and defined, and that provides guidance on how the project scope will be managed and controlled by the project management team. It is contained in or is a subsidiary plan of the project management plan. The project scope management plan can be informal and broadly framed, or formal and highly detailed, based on the needs of the project.

Project Scope Statement   The narrative description of the project scope, including major deliverables, project objectives, project assumptions, project constraints, and a statement of work, that provides a documented basis for making future project decisions and for confirming or developing a common understanding of project scope among the stakeholders. A statement of what needs to be accomplished.

Project Summary Work Breakdown Structure (PSWBS)   A work breakdown structure for the project that is only developed down to the subproject level of detail within some legs of the WBS, and where the detail of those subprojects are provided by use of contract work breakdown structures.

Project Team   All the project team members, including the project management team, the project manager and, for some projects, the project sponsor.

Project Team Directory   A documented list of project team members, their project roles, and communication information.

Project Team Members   The persons who report either directly or indirectly to the project manager, and who are responsible for performing project work, as a regular part of their assigned duties.

Project Organization   Any organizational structure in which the project manager has full authority to assign priorities, apply resources, and direct the work of persons assigned to the project.

Qualitative Risk Analysis   The process of prioritizing risks for subsequent further analysis or action by assessing and combining their probability of occurrence and impact.

Quality   The degree to which a set of inherent characteristics fulfills requirements.

Quality Management Plan   The quality management plan describes how the project management team will implement the performing organization's quality policy. The quality management plan is a component or a subsidiary plan of the project management plan. The quality management plan may be formal or informal, highly detailed, or broadly framed, based on the requirements of the project.

Quality Planning   The process of identifying which quality standards are relevant to the project and determining how to satisfy them.

Quantitative Risk Analysis   The process of numerically analyzing the effect on overall project objectives of identified risks.

Regulation   Requirements imposed by a governmental body. These requirements can establish product, process, or service characteristics—including applicable administrative provisions—that have government-mandated compliance.

Reliability   The probability of a product performing its intended function under specific conditions for a given period of time.

Remaining Duration   The time in calendar units, between the data date of the project schedule and the finish date of a schedule activity that has an actual start date. This represents the time needed to complete a schedule activity where the work is in progress.

Request for Information   A type of procurement document whereby the buyer requests a potential seller to provide various pieces of information related to a product or service or seller capability.

Request for Proposal (RFP)   A type of procurement document used to request proposals from prospective sellers of products or services. In some application areas, it may have a narrower or more specific meaning

Request for Quotation (RFQ)   A type of procurement document used to request price quotations from prospective sellers of common or standard products or services. Sometimes used in place of request for proposal and in some application areas, it may have a narrower or more specific meaning.

Request Seller Responses   The process of obtaining information, quotations, bids, offers, or proposals, as appropriate.

Requested Change   A formally documented change request that is submitted for approval to the integrated change control process. Contrast with **Approved Change Request**.

Requirement   A condition or capability that must be met or possessed by a system, product, service, result, or component to satisfy a contract, standard, specification, or other formally imposed documents. Requirements include the quantified and documented needs, wants, and expectations of the sponsor, customer, and other stakeholders.

Reserve   A provision in the project management plan to mitigate cost and/or schedule risk. Often used with a modifier (e.g., management reserve and contingency reserve) to provide further detail on what types of risk are meant to be mitigated.

Reserve Analysis   An analytical technique to determine the essential features and relationships of components in the project management plan to establish a reserve for the schedule duration, budget, estimated cost, or funds for a project.

Residual Risk   A risk that remains after risk responses have been implemented.

Resource   Skilled human resources (specific disciplines either individually or in crews or teams), equipment, services, supplies, commodities, materiel, budgets, or funds.

Resource Breakdown Structure (RBS)   A hierarchical structure of resources by resource category and resource type used in resource leveling schedules and to develop resource-limited schedules, and may be used to identify and analyze project human resource assignments.

Resource Calendar   A calendar of working days and nonworking days that determines those dates on which each specific resource is idle or can be active. Typically defines resource specific holidays and resource availability periods. See also **Project Calendar**.

Resource-Constrained Schedule   See **Resource-Limited Schedule**.

Resource Histogram   A bar chart showing the amount of time that a resource is scheduled to work over a series of time periods. Resource availability may be depicted as a line for comparison purposes. Contrasting bars may show actual amounts of resource used as the project progresses.

Resource Leveling   Any form of schedule network analysis in which scheduling decisions (start and finish dates) are driven by resource constraints (e.g., limited resource availability or difficult-to-manage changes in resource availability levels).

Resource-Limited Schedule   A project schedule whose schedule activity, scheduled start dates and scheduled finish dates reflect expected resource availability. A resource-limited schedule does not have any early or late start or finish dates. The resource-limited schedule total float is determined by calculating the difference between the critical path method late finish date and the resource-limited scheduled finish date. Sometimes called resource-constrained schedule. See also **Resource Leveling**.

Resource Planning   See **Activity Resource Estimating**.

Responsibility Matrix   A structure that relates the project organizational breakdown structure to the work breakdown structure to help ensure that each component of the project's scope of work is assigned to a responsible person.

Result   An output from performing project management processes and activities. Results include outcomes (e.g., integrated systems, revised process, restructured organization, tests, trained personnel, etc.) and documents (e.g., policies, plans, studies, procedures, specifications, reports, etc.).

Retainage   A portion of a contract payment that is withheld until contract completion to ensure full performance of the contract terms.

Rework   Action taken to bring a defective or nonconforming component into compliance with requirements or specifications.

Risk   An uncertain event or condition that, if it occurs, has a positive or negative effect on a project's objectives. See also **Risk Category** and **Risk Breakdown Structure**.

Risk Acceptance   A risk-response-planning technique that indicates that the project team has decided not to change the project management plan to deal with a risk, or is unable to identify any other suitable response strategy.

Risk Avoidance   A risk response-planning technique for a threat that creates changes to the project management plan that are meant to either eliminate the risk or to protect the project objectives from its impact. Generally, risk avoidance involves relaxing the time, cost, scope, or quality objectives.

Risk Breakdown Structure (RBS)   A hierarchically organized depiction of the identified project risks arranged by risk category and subcategory that identifies the various areas and causes of potential risks. The risk breakdown structure is often tailored to specific project types.

Risk Category   A group of potential causes of risk. Risk causes may be grouped into categories such as technical, external, organizational, environmental, or project management. A category may include subcategories such as technical maturity, weather, or aggressive estimating. See also **Risk Breakdown Structure**.

Risk Database   A repository that provides for collection, maintenance, and analysis of data gathered and used in the risk management processes.

Risk Identification   The process of determining which risks might affect the project and documenting their characteristics.

Risk Management Plan   The document describing how project risk management will be structured and performed on the project. It is contained in or is a subsidiary plan of the project management plan. The risk management plan can be informal and broadly framed, or formal and highly detailed, based on the needs of the project. Information in the risk management plan varies by application area and project size. The risk management plan is different from the risk register that contains the list of project risks, the results of risk analysis, and the risk responses.

Risk Management Planning   The process of deciding how to approach, plan, and execute risk management activities for a project.

Risk Mitigation   A risk-response-planning technique associated with threats that reduces the probability of occurrence or impact of a risk to below an acceptable threshold.

Risk Monitoring and Control   The process of tracking identified risks, monitoring residual risks, identifying new risks, executing risk response plans, and evaluating their effectiveness throughout the project life cycle.

Risk Register   The document containing the results of the qualitative risk, analysis, quantitative risk analysis, and risk response planning. The risk register details all identified risks, including description, category, cause, probability of occurring, impacts on objectives, proposed responses, owners, and current status. The risk register is a component of the project management plan.

Risk Response Planning   The process of developing options and actions to enhance opportunities and to reduce threats to project objectives.

Risk Transference   A risk-response-planning technique that shifts the impact of a threat to a third party, together with ownership of the response.

Role   A defined function to be performed by a project team member, such as testing, filing, inspecting, and coding.

Rolling Wave Planning   A form of progressive elaboration planning in which the work to be accomplished in the near term is planned in detail at a low level of the work breakdown structure, while the work far in the future is planned at a relatively high level of the work breakdown structure, but the detailed planning of the work to be performed within another one or two periods in the near future is done as work is being completed during the current period.

Root Cause Analysis   An analytical technique used to determine the basic underlying reason that causes a variance or a defect or a risk. A root cause may underlie more than one variance or defect or risk.

Schedule   See **Project Schedule** and **Schedule Model**.

Schedule Activity   A discrete scheduled component of work performed during the course of a project. A schedule activity normally has an estimated duration, an estimated cost, and estimated resource requirements. Schedule activities are connected to other schedule activities or schedule milestones with logical relationships, and are decomposed from work packages.

Schedule Analysis   See **Schedule Network Analysis**.

Schedule Compression   Shortening the project schedule duration without reducing the project scope. See also **Crashing** and **Fast Tracking**.

Schedule Control   The process of controlling changes to the project schedule.

Schedule Development   The process of analyzing schedule activity sequences, schedule activity durations, resource requirements, and schedule constraints to create the project schedule.

Schedule Management Plan   The document that establishes criteria and the activities for developing and controlling the project schedule. It is contained in or is a subsidiary of the project management plan. The schedule management plan may be formal or informal, highly detailed or broadly framed, based on the needs of the project.

Schedule Milestone   A significant event in the project schedule, such as an event restraining future work or marking the completion of a major deliverable. A schedule milestone has zero duration. Sometimes called a milestone activity. See also **Milestone**.

Schedule Model   A model used in conjunction with manual methods or project management software to perform schedule network analysis to generate the project schedule for use in managing the execution of a project. See also **Project Schedule**.

Schedule Network Analysis   The technique of identifying early and late start dates, as well as early and late finish dates, for the uncompleted portions of project schedule activities. See also **Critical Path Method, Critical Chain Method, What-If Analysis,** and **Resource Leveling**.

Schedule Performance Index (SPI)   A measure of schedule efficiency on a project. It is the ratio of earned value (EV) to planned value (PV), that is, SPI = EV/PV. An SPI value equal to or greater than one indicates a favorable condition; and a value of less than 1 indicates an unfavorable condition. See also **earned value management**.

Schedule Variance (SV)   A measure of schedule performance on a project. It is the algebraic difference between the earned value (EV) and the planned value (PV), that is, $SV = EV - PV$. See also **Earned Value Management**.

Scheduled Finish Date   The point in time at which work was scheduled to finish on a schedule activity. The scheduled finish date is normally within the range of dates delimited by the early finish date and the late finish date. It may reflect resource leveling of scarce resources. Sometimes called planned finish date.

Scheduled Start Date   The point in time at which work was scheduled to start on a schedule activity. The scheduled start date is normally within the range of dates delimited by the early start date and the late start date. It may reflect resource leveling of scarce resources. Sometimes called planned start date.

Scope   The sum of the products, services, and results to be provided as a project. See also **Project Scope** and **Product Scope**.

Scope Baseline   See **Baseline**.

Scope Change   Any change to the project scope. A scope change almost always requires an adjustment to the project cost or schedule.

Scope Control   The process of controlling changes to the project scope.

Scope Creep   Adding features and functionality (project scope) without addressing the effects on time, costs, and resources, or without customer approval.

Scope Definition   The process of developing a detailed project scope statement as the basis for future project decisions.

Scope Verification   The process of formalizing acceptance of the completed project deliverable.

S-Curve   Graphic display of cumulative costs, labor hours, percentage of work, or other quantities, plotted against time. The name derives from the S-like shape of the curve (flatter at the beginning and end, steeper in the middle) produced on & project that starts slowly, accelerates, and then tails off. Also a term for the cumulative likelihood distribution that is a result of a simulation, a tool of quantitative risk analysis.

Secondary Risk   A risk that arises as a direct result of implementing a risk response.

Select Sellers   The process of reviewing offers, choosing from among potential sellers, and negotiating a written contract with a seller.

Seller   A provider or supplier of products, services, or results to an organization.

Sensitivity Analysis   A quantitative risk analysis and modeling technique used to help determine which risks have the most potential impact on the project. It examines the extent to which the uncertainty of each project element affects the objective being examined when all other uncertain elements are held at their baseline values. The typical display of results is in the form of a tornado diagram.

Service   Useful work performed that does not produce a tangible product or result, such as performing any of the business functions supporting production or distribution. Contrast with **Product** and **Result**. See also **Deliverable**.

Simulation   A simulation uses a. project model that translates the uncertainties specified at a detailed level into their potential impact on objectives that are expressed at the level of the total project. Project simulations use computer models and estimates of risk, usually expressed as a probability distribution of possible costs or durations at a detailed work level, and are typically performed using Monte Carlo analysis.

Skill   Ability to use knowledge, a developed aptitude, and/or a capability to effectively and readily execute or perform an activity.

Slack   See **Total Float** and **Free Float**.

Special Cause   A source of variation that is not inherent in the system, is not predictable, and is intermittent. It can be assigned to a defect in the system. On a control chart, points beyond the control limits, or nonrandom patterns within the control limits, indicate it. Also referred to as assignable cause. Contrast with **Common Cause**.

Specification   A document that specifies, in a complete, precise, verifiable manner, the requirements, design, behavior, or other characteristics of a system, component, product, result, or service and, often, the procedures for determining whether these provisions have been satisfied. Examples are requirement specification, design specification, product specification, and test specification.

Specification Limits   The area, on either side of the centerline, or mean, of data plotted on a control chart that meets the customer's requirements for a product or service. This area may be greater than or less than the area defined by the control limits. See also **Control Limits**.

Sponsor   The person or group that provides the financial resources, in cash or in kind, for the project.

Staffing Management Plan   The document that describes when and how human resource requirements will be met. It is contained in, or is a subsidiary plan of, the project management plan. The staffing management plan can be informal and broadly framed, or formal and highly detailed, based on the needs of the project. Information in the staffing management plan varies by application area and project size.

Stakeholder   Persons and organizations such as customers, sponsors, performing organization and the public, that are actively involved in the project, or whose interests may be positively or negatively affected by execution or completion of the project. They may also exert influence over the project and its deliverable.

Standard   A document established by consensus and approved by a recognized body that provides, for common and repeated use, rules, guidelines, or characteristics for activities or their results aimed at the achievement of the optimum degree of order in a given context.

Start Date   A point in time associated with a schedule activity's start, usually qualified by one of the following: actual, planned, estimated, scheduled, early, late, target, baseline, or current.

Start-to-Finish   The logical relationship where completion of the successor schedule activity is dependent on the initiation of the predecessor schedule activity. See also **Logical Relationship**.

Start-to-Start   The logical relationship where initiation of the work of the successor schedule activity depends on the initiation of the work of the predecessor schedule activity. See also **Logical Relationship**.

Statement of Work (SOW)   A narrative description of products, services, or results to be supplied.

Subnetwork A subdivision (fragment) of a project schedule network diagram, usually representing a subproject or a work package. Often used to illustrate or study some potential or proposed schedule condition, such as changes in preferential schedule logic or project scope.

Subphase A subdivision of a phase.

Subproject A smaller portion of the overall project created when a project is subdivided into more manageable components or pieces. Subprojects are usually represented in the work breakdown structure. A subproject can be referred to as a project, managed as a project, and acquired from a seller. May be referred to as a subnetwork in a project schedule network diagram.

Successor See **Successor Activity**.

Successor Activity The schedule activity that follows and predecessor activity, as determined by their logical relationship.

Summary Activity A group of related schedule activities aggregated at some summary level, and displayed/reported as a single activity at that summary level. See also **Subproject and Subnetwork**.

SWOT Analysis (Strengths, Weaknesses, Opportunities, and Threats Analysis) This information gathering technique examines the project from the perspective of each project's strengths, weaknesses, opportunities, and threats to increase the breadth of the risks considered by risk management.

System An integrated set of regularly interacting or interdependent components created to accomplish a defined objective, with defined and maintained relationships among its components, and the whole producing or operating better than the simple sum of its components. Systems may be either physically process based or management process based, or more commonly a combination of both. Systems for project management are composed of project management processes, techniques, methodologies, and tools operated by the project management team.

Target Completion Date An imposed date that constrains or otherwise modifies the schedule network analysis.

Target Finish Date The date that work is planned (targeted) to finish on a schedule activity.

Target Schedule A schedule adopted for comparison purposes during schedule network, analysis, which can be different from the baseline schedule. See also **Baseline**.

Target Start Date The date that work is planned (targeted) to start on a schedule activity.

Task A term for work, whose meaning and placement within a structured plan for project work varies by the application area, industry, and brand of project management software.

Technical Performance Measurement A performance measurement technique that compares technical accomplishments during project execution with the project management plan's schedule of planned technical achievements. It may use key technical parameters of the product produced by the project as a quality metric. The achieved metric values are part of the work performance information.

Template A partially complete document in a predefined format that provides a defined structure for collecting, organizing, and presenting information and data.

Templates are often based upon documents created during prior projects. Templates can reduce the effort needed to perform work and increase the consistency of results.

Threat   A condition or situation unfavorable to the project, a negative set of circumstances, a negative set of events, a risk that will have a negative impact on a project objective if it occurs, or a possibility for negative changes. Contrast with **Opportunity**.

Three-Point Estimate   An analytical technique that uses three cost or duration estimates to represent the optimistic, most likely, and pessimistic scenarios. This technique is applied to improve the accuracy of the estimates of cost or duration when the underlying activity or cost component is uncertain.

Threshold   A cost, time, quality, technical, or resource value used as a parameter, and which may be included in product specifications. Crossing the threshold should trigger some action, such as generating an exception report.

Time and Material (T&M) Contract   A type of contract that is a hybrid contractual arrangement containing aspects of both cost reimbursable and fixed-price contracts. Time and material contracts resemble cost-reimbursable-type arrangements in that they have no definitive end, because the full value of the arrangement is not defined at the time of the award. Thus, time and material contracts can grow in contract value as if they were cost-reimbursable-type arrangements. Conversely, time and material arrangements can also resemble fixed-price arrangements. For example, the unit rates are preset by the buyer and seller, when both parties agree on the rates for the category of senior engineers.

Time-Now Date   See **Data Date**.

Time-Scaled Schedule Network Diagram   Any project schedule network diagram drawn in such a way that the positioning and length of the schedule activity represents its duration. Essentially, it is a bar chart that includes schedule network logic.

Total Float   The total amount of time that a schedule activity may be delayed from its early start date without delaying the project finish date or violating a schedule constraint. Calculated using the Critical Path Method technique and determining the difference between the early finish dates and late finish dates. See also **Free Float**.

Total Quality Management (TQM)   A common approach to implementing a quality improvement program within an organization.

Trend Analysis   An analytical technique that uses mathematical models to forecast future outcomes based on historical results. It is a method of determining the variance from a baseline of a budget, cost, schedule, or scope parameter by using prior progress reporting periods' data and projecting how much that parameter's variance from baseline might be at some future point in the project if no changes are made in executing the project.

Triggers   Indications that a risk has occurred or is about to occur. Triggers may be discovered in the risk identification process and watched in the risk monitoring and control process. Triggers are sometimes called risk symptoms or warning signs.

Triple Constraint   A framework for evaluating competing demands. The triple constraint is often depicted as a triangle in which one of the sides or corners represents one of the parameters being managed by the project team.

User   The person or organization that will use the project's product or service. See also Customer.

Validation   The technique of evaluating a component or product during or at the end of a phase or project to ensure it complies with the specified requirements. Contrast with Verification.

Value Engineering (VE)   A creative approach used to optimize project life cycle costs, save time, increase profits, improve quality, expand market share, solve problems, and/or use resources more effectively.

Variance   A quantifiable deviation, departure, or divergence away from a known baseline or expected value.

Variance Analysis   A method for resolving the total variance in the set of scope, cost, and schedule variables into specific component variances; that are associated with defined factors affecting the scope, cost, and schedule variables.

Verification   The technique of evaluating a component or product at the end of a phase or project to assure or confirm it satisfies the conditions imposed. Contrast with **Validation**.

Virtual Team   A group of persons with a shared objective who fulfill their roles with little or no time spent meeting face to face. Various forms of technology are often used to facilitate communication among team members. Virtual teams can be comprised of persons separated by great distances.

Voice of the Customer   A planning technique used to provide products, services, and results that truly reflect customer requirements by translating those customer requirements into the appropriate technical requirements for each phase of project product development.

War Room   A room used for project conferences and planning, often displaying charts of cost, schedule status, and other key project data.

Work   Sustained physical or mental effort, exertion, or exercise of skill to overcome obstacles and achieve an objective.

Work Authorization   A permission and direction, typically written, to begin work on a specific schedule activity or work package or control account. It is a method for sanctioning project work to ensure that the work is done by the identified organization, at the right time, and in the proper sequence.

Work Authorization System   A subsystem of the overall project management system. It is a collection of formal documented procedures that defines haw project work will be authorized (committed) to ensure that the work is done by the identified organization, at the right time, and in the proper sequence. It includes the steps, documents, tracking system, and defined approval levels needed to issue work authorizations.

Work Breakdown Structure (WBS)   A deliverable-oriented hierarchical decomposition of the work, to be executed by the project team to accomplish the project objectives and create the required deliverables. It organizes and defines the total scope of the project. Each descending level represents an increasingly detailed definition of the project work. The WBS is decomposed into work packages. The deliverable orientation of the hierarchy includes both internal and external deliverables. See also work package, control account, contract work, breakdown structure, and project summary work, breakdown structure.

Work Breakdown Structure Component   An entry in the work breakdown structure that can be at any level.

Work Breakdown Structure Dictionary   A document that describes each component in the work, breakdown structure (WBS). For each WBS component, the WBS dictionary includes a brief definition of the scope or statement of work, defined deliverables, a list of associated activities, and a list of milestones. Other information may include: responsible organization, start and end dates, resources required, an estimate of cost, charge number, contract information, quality requirements, and technical references to facilitate performance of the work.

Work Item   See **Activity and Schedule Activity**.

Work Package   A deliverable or project work component at the lowest level of each branch of the work breakdown structure. The work package includes the schedule activities and schedule milestones required to complete the work package deliverable or project work component. See also Control Account.

Work Performance Information   Information and data on the status of the project schedule activities being performed to accomplish the project work, collected as part of the direct and manage project execution processes. Information includes status of deliverable, implementation status for change requests, corrective actions, preventive actions, and defect repairs, forecasted estimates to complete, reported percent of work physically completed; achieved value of technical performance measures, and start and finish dates of schedule activities.

Workaround   A response to a negative risk that has occurred. Distinguished from contingency plan in that a workaround is not planned in advance of the occurrence of the risk event.

Appendix B: Project Acronyms

A&E	architecture and engineering
AACE	American Association of Cost Engineers
ABC	activity-based costing
ACO	administrative contracting officer
ACV	at-completion variance
ACWP	actual cost of work performed
ADP	automated data processing
ADPE	automated data processing equipment
ADR	Arrow Diagramming Method
AF	award fee
AFR	air force regulation
AGE	auxiliary ground equipment
AHP	analytical hierarchy process
AIS	automated information system
ANSI	American National Standards Institute
AOA	activity on arrow
AON	activity on node
APR	Acquisition Plan Review
AQL	acceptable quality level
AR	Acceptance Review
ARB	acquisition review board
ARC	Appraisal Requirements for CMMI
ARO	after receipt of order
ASAPM	American Society for the Advancement of Project Management
ASCR	Annual System Certification Review
AT	acceptance test
ATE	automatic test equipment
ATP	acceptance test procedure
AUW	authorized unpriced work
B&P	bid and proposal funds
BAA	broad agency announcement
BAC	budget at completion
BAFO	best and final offer
BCE	baseline cost estimate
BCWP	budgeted cost of work performed
BCWS	budgeted cost of work scheduled
BIT	built-in test

BITE	built-in test equipment
BNB	bid/no bid
BOA	basic ordering agreement
BOE	basis of estimate
BOM	bill of material
BPA	blanket purchase agreement
BTW	by the way
BY (1)	base year
BY (2)	budget year
C/SCSC	cost/schedule control system criteria
C/SSR	cost/schedule status report
CA	contract administrator
CAAS	contracted advisory and assistance services
CAC	cost at completion
CAD	computer-aided design
CADM	computer-aided document management
CAIV	cost as an independent variable
CAM (1)	computer aided manufacturing
CAM (2)	cost account manager
CAR	Contract Acceptance Review
CAS	cost accounting standards
CASE (1)	computer-aided software engineering
CASE (2)	computer-aided systems engineering
CAT	computer-aided testing
CBD	Commerce Business Daily
CBJ	congressional budget justification
CBJR	Congressional Budget Justification Review
CCA	change control authority
CCB (1)	change control board
CCB (2)	configuration control board
CCN	contract change notice
CCO	contract change order
CCP	contract change proposal
CDCG	Contract Data Classification Guide
CDD	Concept Definition Document
CDR	Critical Design Review
CDRL (1)	contract data requirements lists
CDRL (2)	contract documentation requirements list
CEO	chief executive officer
CET	cost evaluation team
CFE	contractor furnished equipment
CFSR	contract funds status report
CI (1)	configuration item
CI (2)	continuous improvement
CIAR	Configuration Item Acceptance Review
CICA	Competition In Contracting Act Of 1984

CID	commercial item description
CIR (1)	Contract Implementation Review
CIR (2)	contract inspection report
CIT	component integration and test
CITR	Configuration Item Test Readiness
CLIN	contract line item numbers
CM	configuration management
CMM	Capability Maturity Model
CMMI	Capability Maturity Model Integration
CMO	configuration management officer
CMSP	contractor management systems evaluation program
CO (1)	change order
CO (2)	contracting officer
COCOMO	constructive cost model
CONOPS (1)	concept of operations
CONOPS (2)	user concept of operations
CONOPS (3)	system concept of operations
COR	contracting officer's representative
COTR	contracting officer's technical representative
COTS	commercial off-the-shelf
COW	cards on the wall planning
CPA	certified public accountant
CPAF	cost plus award fixed
CPC	computer program component
CPCI	computer program configuration item
CPFF	cost plus fixed fee
CPI (1)	continuous process improvement
CPI (2)	Cost Performance Index
CPIF	cost plus incentive fee
CPO	contractor project office
CPR	cost performance report
CPU	central processing unit
CPVR	Construction Performance Verification Review
CQI	continuous quality improvement
CRADA	cooperative research and development agreement
CRWG	computer resources working group
CSC	computer software component
CSCI	computer software configuration item
CSE	chief systems engineer
CSOM	computer system operators manual
CSSR	contract system status report
CSU	computer software unit
CTC (1)	collaborate to consensus
CTC (2)	contract target cost
CTC (3)	cost to complete
CTP	contract target price

CV	cost variance
CWBS	contract work breakdown structure
CY	calendar year
DA&R	decomposition analysis and resolution
DAR	Deactivation Approval Review
DARPA	Defense Advanced Research Projects Agency
DCAA	Defense Contract Audit Agency
DCAS	Defense Contract Administration Service
DCN	documentation change notice
DCR	Design Concept Review
DD 250	Department of Defense form 250
DDT&E	design, development, test, and evaluation
DID	data item description
DLA	Defense Logistics Agency
DMO	documentation management officer
DP	data processing
DPAS	Defense Priorities And Allocation System
DPRO	Defense Plant Representative Office
DR	discrepancy report
DRD	documentation requirements description
DRR	Development Readiness Review
DSMC	Defense Systems Management College
DT&E	development test and evaluation
DTC	design to cost
DTS	design to schedule
DVR	Documentation Verification Review
EAC	Estimate At Completion
ECCM	electronic counter-countermeasures
ECD	estimated completion date
ECM	electronic countermeasures
ECN	engineering change notice
ECP	engineering change proposal
ECR	engineering change request
EDM	engineering development model
EI	end item
EMC	electromagnetic compatibility
EMI	electromagnetic interference
EO	engineering order
ERB	engineering review broad
ESS	environmental stress screening
ETC	estimate to complete
ETR	estimated time to repair
EV	earned value
EW	electronic warfare
FA	first article
FAR	Federal Acquisition Regulations

FARA	Federal Acquisition Reform Act
FASA	Federal Acquisition Streamlining Act
FAT (1)	factory acceptance test
FAT (2)	first article test
FCA	functional configuration audit
FCCM	facilities capital cost of money
FCR (1)	Final Contract Review
FCR (2)	Facility Contract Review
FDR	final Design Review
FFBD	functional flow block diagram
FFP	Firm Fixed Price contract
FFRDC	Federally Funded Research and Development Center
FMEA	failure mode and effects analysis
FMECA	failure mode, effects, and criticality analysis
FOC	full operational capability
FOIA	Freedom of Information Act
FOM	figure of merit
FP	fixed-price contract
FPAF	fixed-price award fee
FPIF	fixed-price incentive fee
FPR (1)	Final Proposal Review
FPR (2)	fixed-price redeterminable
FPVR	Facility Performance Verification Review
FQR	Formal Qualification Review
FQT	formal qualification testing
FRB	failure review board
FRR	Facility Readiness Review
FSOW	facility scope of work
FTRR	Facility Test Readiness Review
FY	fiscal year
G&A	general and administrative costs
GAO	General Accounting Office
GAS	general accounting system
GFE	government-furnished equipment
GFF	government-furnished facilities
GFI	government-furnished information
GFM	government-furnished material
GFP	government-furnished property
GOCO	government-owned, contractor operated
GOGO	government-owned, government operated
GOTS	government off-the-shelf
GPO	government project office
GSA	General Services Administration
GSE	ground support equipment
HAC	House Appropriations Committee
HCI	human computer interface

HQ	headquarters
HW	hardware
HWCI	hardware configuration item
IAW	in accordance with
ICD	Interface Control Document
ICP	Interface Control Plan
ICWG	interface control working group
ID (1)	identifier
ID (2)	independent development
ID (3)	indefinite delivery contract
IDD	interface design document
IDEFO	Integrated Definition for Functional Modeling
IE	information engineering
IEEE	Institute of Electrical and Electronics Engineers
IFB	invitation for bid
IG	inspector general
IGCE	independent government cost estimate
ILS	integrated logistics support
INCOSE	International Council on Systems Engineering
INI	interest/no interest
IOC	initial operational capability
IPT (1)	integrated product teams
IPT (2)	integrated project teams
IQ	indefinite quantity
IR&D	independent research and development
IRR	internal rate of return
IRS	interface requirements specification
IS	interface specification
ISCO	integrated schedule commitment
ISO	International Organization for Standardization
IV&V (1)	independent verification and validation
IV&V (2)	integration verification and validation
L/H	labor hour contract
LAN	local area network
LCC	life cycle cost
LOB	line of business
LOC (1)	lines of code
LOC (2)	logistics operations center
LOE	level of effort
LOI	letter of intent
LRIP	low-rate initial production
MBO	management by objectives
MBWA	management by walking around
MDT	mean down time
MIV	management information center
MIL-SPEC	military specification

MIL-STD	military standard
MIPR	Military Interdepartmental Purchase Request
MIS	management information system
MOA	memorandum of agreement
MOE	measure of effectiveness
MOP	measure of performance
MOU	memorandum of understanding
MPS	master project schedule
MR	management reverse
MRB	material review board
MRP (1)	manufacturing resource planning
MRP (2)	material resource planning
MTBF	mean time between failures
MTTR	mean time to repair
MYP	multiyear procurement
N/A	not applicable
NBV	net book value
NC	numerical control
NCR	nonconformance report
NDI	nondevelopment item
NIH	not invented here
NLT	no later than
NMT	not more than
NPV	net present value
NTE	not to exceed
O&M	operations and maintenance
OAR	Operational Acceptance Review
OBS	organizational breakdown structure
ODC	other direct costs
OGC	Office of General Council
OH	overhead
OJT	on-the-job training
OMB	Office of Management and Budget
OOA	object-oriented analysis
OOD	object-oriented design
ORC	operational readiness certificate
ORD	Operational Requirements Document
ORR	Operational Readiness Review
OSHA	Occupational Safety and Health Administration
OT&E	operational test and evaluation
OVR	Operational Validation Review
PA	product assurance
PAR	Product Acceptance Review
PBS	product breakdown structure
PC (1)	personal computer
PC (2)	project cycle

PCA	physical configuration audit
PCCB	project configuration control board
PCO (1)	procuring contracting officer
PCO (2)	principal contracting officer
PCR	Project Completion Review
PDP	previously developed products
PDR	Preliminary Design Review
PERT	Project Evaluation Review Technique
PET	proposal evaluation teams
PIP	product improvement plan
PIR (1)	Project Implementation Review
PIR (2)	Project Initiation Review
PL	public law
PM (1)	program manager
PM (2)	project manager
PM&P	parts, material, and processes
PMB	performance measurement baseline
PMBOK®	Project Management Body of Knowledge
PMI®	Project Management Institute
PMP®	Project Management Professional
PMS	performance measurement system
PNP	pursue/no pursue
POC	point of contact
POM	program management memorandum
PPI	proposal preparation instructions
PPL	project products list
PPLFS	projects products list fact sheets
PPPI	preplanned product improvement
PPR	Project Plans Review
PRB	project review board
PRICE	Program Review Information for Costing and Estimating
PRR	Production Readiness Review
PSR	Project Specification Review
PWAA	Project Work Authorizing Agreement
PY	prior year
QA	quality assurance
QAR	Qualification Acceptance Review
QC	quality control
QFD	quality function deployment
QRC	quick reaction capability
R&D	research and development
RAD	rapid application development
RAM	random access memory
RDT&E	research, development, test, and evaluation
RFC	request for change
RFI	request for information

RFP	request for proposal
RFQ	request for quotation
RIF	reduction in force
ROI	return on investment
ROM (1)	rough order of magnitude
ROM (2)	read-only memory
RTM	requirements traceability matrix
RTVM	requirement traceability and verification matrix
RVM	requirements verification matrix
S/C	subcontract
SAP	system acquisition plan
SAR	System Acceptance Review
SBA	Small Business Administration
SCA	subcontract administrator
SCAMPI	Standard CMMI Appraisal Method for Process Improvement
SCE^{sm}	Software Capability Evaluation
SCN	specification change notice
SCR	System Concept Review
SDD	Software Design Document
SDF	software development file
SDL	software development library
SDP	software development plan
SDR	System Design Review
SDRL	subcontract documentation requirements list
SEB	source evaluation board
SEI	Software Engineering Institute
SEI&T	systems engineering, integration, and test
SEMP	systems engineering management plan
SETA	systems engineering and technical assistance
SI	system integrator
SMAP	software management and assurance program
SMT	subcontract management team
SOP	standard operating procedure
SOW	statement of work
SPI	Schedule Performance Index
SPM	software programmers manual
SPO	system project office
SPR	software problem report
SPS	software product specification
SQA	software quality assurance
SRD	system requirements document
SRR	System Requirements Review
SRS	software requirements specification
SSA	source selection authority
SSAC	source selection advisory council
SSAR	Source Selection Authorization Review

SSDD	system/segment design document
SSE	software support environment
SSEB	source selection evaluation board
SSIR	Source Selection Initiation Review
SSM	software sizing model
SSO	source selection official
SSP	source selection plan
SSR	Software Specification Review
SSS	system/segment specification
STE	special test equipment
STP	System Test Plan
STR	software test report
STRR	System Test Readiness Review
SUM	software users manual
SV	schedule variance
SW	software
SWAG	scientific wild anatomical guess
T&E	test and evaluation
T&M	time and materials contract
TAAF	test, analyze, and fix
TBD	to be determined
TBR	to be resolved
TBS	to be supplied
TCPI	to complete performance index
TD	test director
TEM	technical exchange meeting
TET	technical evaluation team
TIM	technical interchange meeting
TM	technical manual
TOC	theory of constraints
TP	test procedures
TPM	technical performance measurement
TQM	Total Quality Management
TR (1)	time remaining
TR (2)	test report
TRR	Test Readiness Review
TTC	time to complete
UAR	User Acceptance Review
UB	undistributed budget
URR	User Readiness Review
VA&R	verification analysis and resolution
VAC	variance at completion
VDD	version description document
VE	value engineering
VECP	value engineering change proposal
VRIC	vendor request for information or change

VV&T	verification, validation, and test
WAN	wide area network
WBS	work breakdown structure
W-Mgt	W Theory Management
WO/WA	work order/work authorization
WP	work packages
WR	work remaining
X-Mgmt	X Theory (or authoritative) Management
Y-Mgmt	Y Theory (or supportive) Management
Z-Mgmt	Z Theory (or participative) Management

Index